Bio-Inspired Fault-Tolerant Algorithms for Network-on-Chip

Bio-Inspired Fault-Tolerant Algorithms for Network-on-Chip

By
Muhammad Athar Javed Sethi

CRC Press is an imprint of the
Taylor & Francis Group, an **informa** business

CRC Press
Taylor & Francis Group
52 Vanderbilt Avenue,
New York, NY 10017

© 2020 by Taylor & Francis Group, LLC

CRC Press is an imprint of Taylor & Francis Group, an Informa business

No claim to original U.S. Government works

Printed on acid-free paper

International Standard Book Number-13: 9780367425906 (Hardback)

This book contains information obtained from authentic and highly regarded sources. Reasonable efforts have been made to publish reliable data and information, but the author and publisher cannot assume responsibility for the validity of all materials or the consequences of their use. The authors and publishers have attempted to trace the copyright holders of all material reproduced in this publication and apologize to copyright holders if permission to publish in this form has not been obtained. If any copyright material has not been acknowledged please write and let us know so we may rectify in any future reprint.

Except as permitted under U.S. Copyright Law, no part of this book may be reprinted, reproduced, transmitted, or utilized in any form by any electronic, mechanical, or other means, now known or hereafter invented, including photocopying, microfilming, and recording, or in any information storage or retrieval system, without written permission from the publishers.

For permission to photocopy or use material electronically from this work, please access www. copyright.com (www.copyright.com/) or contact the Copyright Clearance Center, Inc. (CCC), 222 Rosewood Drive, Danvers, MA 01923, 978–750–8400. CCC is a not-for-profit organization that provides licenses and registration for a variety of users. For organizations that have been granted a photocopy license by the CCC, a separate system of payment has been arranged.

Trademark Notice: Product or corporate names may be trademarks or registered trademarks, and are used only for identification and explanation without intent to infringe.

Library of Congress Cataloging-in-Publication Data
A catalog record for this book has been requested

Visit the Taylor & Francis Web site at
www.taylorandfrancis.com

and the CRC Press Web site at
www.crcpress.com

Dedication

I dedicate this book to my wife, whose constant motivation, support and encouragement have helped me to write the book.

Contents

Preface xi
Author Biography xiii
List of Abbreviations xv
List of Symbols xvii

Chapter 1   Introduction 1

       1.1   Network on Chip 1
       1.2   Fault Tolerance 2
       1.3   Why Use Bio-Inspired Algorithms for Fault Tolerance 4
       1.4   Objectives of Bio-Inspired Fault-Tolerant Algorithms 6
       1.5   Book Contributions 6
       1.6   Organization of the Book 7

Chapter 2   In-Depth Review of Network on Chip 9

       2.1   Link-Sharing Mechanism 9
       2.2   NoC Fault-Tolerant Routing Algorithms 11
       2.3   Switching Techniques 17
       2.4   Buffer Management Techniques 19
       2.5   NoC Evaluation Parameters 20
       2.6   NoC Clocking Mechanism 21
       2.7   NoC Topologies 24
       2.8   Open Source 25
       2.9   NoC Connection Types 26
       2.10 NoC Size 28
       2.11 NoC Implementation Platforms 28
       2.12 NoC Buffering Mechanisms 28
       2.13 NoC PE–Router Interface 29
       2.14 NoC Frequency and Technology 32
       2.15 NoC Area and Power Consumption 33
       2.16 NoC Router Ports and Bus Width 34
       2.17 NoC Year of Proposal, Flit Size and Latency 34
       2.18 Quality of Service 36
       Summary 38

Chapter 3   Bio-Inspired Algorithms and Implementation 49

       3.1   Swarm Intelligence Algorithms 49
       3.2   Ant Colony Optimization 49
       3.3   Artificial Immune System 49

viii Contents

3.4	Firefly Algorithm	49
3.5	Epidemic Spreading	49
3.6	Flower Pollination Algorithm	50
3.7	Artificial Bee Colony Algorithm	50
3.8	Cat Swarm Optimization	50
3.9	Cuckoo Search	50
3.10	Bat Algorithm	51
3.11	Cuttlefish Algorithm	51
3.12	Harris Hawks Optimization	51
3.13	Killer Whale Algorithm	51
3.14	Cobweb Network on Chip Topology	51
3.15	Scalable Bio-Inspired Fault Detection Unit in Network on Chip	51
3.16	Autonomous Error Tolerant Architecture	52
3.17	SpiNNaker Communication	52
3.18	Autonomic Network on Chip Using the Biological Immune System	52
3.19	Fault-Tolerant NoC Using Biological Brain Techniques	52
3.20	Bio-Inspired Online Fault Detection in the NoC Interconnect	52
3.21	Bio-Inspired Self-Aware NoC Fault-Tolerant Routing Algorithm	52

Chapter 4 Bio-Inspired NoC Fault-Tolerant Algorithms 55

4.1	Biological Brain Characteristics	55
4.2	Synaptogenesis	56
4.3	Sprouting	57
4.4	Bio-Inspired NoC Algorithms	57
4.5	Bio-Inspired NoC Framework	59
4.6	Bio-Inspired NoC Network	59
4.7	Bio-Inspired NoC Fault-Tolerant Algorithm	62
4.8	Bio-Inspired BE and GT NoC Algorithm and Architectures	74
Summary		88

Chapter 5 Analysis of Bio-Inspired NoC Fault-Tolerant Algorithms 89

5.1	Research Framework, Design and Parameters	89
5.2	Bio-Inspired NoC Fault-Tolerant Algorithms Analysis Results	91
Summary		106

Contents

Chapter 6 Conclusion and Future Work 109

       6.1   Future Work 110

Appendix A 113

Appendix B 185

Index 191

Preface

The book is about bio-inspired fault-tolerant algorithms for network on chip (NoC). The inspiration for fault tolerance came from the biological brain, as it is highly robust and fault-tolerant itself. Synaptogenesis and sprouting are self-adapting and self-healing mechanisms of the brain being implemented in NoC. The results are quite promising, as these have made the NoC architecture fault-tolerant and reliable in terms of communication. To explain the bio-inspired algorithms efficiently, different NoC architectural parameters are presented in the book. Moreover, different bio-inspired techniques are also presented in the book that represent how nature solves real-world problems.

Author Biography

Dr. **Muhammad Athar Javed Sethi** is an assistant professor at the Department of Computer Systems Engineering, University of Engineering and Technology (UET), Peshawar, Pakistan. He received a Bachelor of Sciences in computer information systems engineering (honors) from UET Peshawar, Pakistan, in 2004. He received a Master of Sciences in computer systems engineering from the same university in 2008. He completed his PhD at Universiti Teknologi PETRONAS (UTP), Malaysia, in the Department of Electrical and Electronic Engineering in 2016. He was able to get different scholarships to pursue his studies. Dr. Sethi received the Institute of Electrical and Electronics Engineers (IEEE) award for the best student paper in 2013 and the most downloaded paper award from Elsevier in 2016 and 2017 for two different papers. His research interests include network on chip (NoC), interconnection networks, computer architecture and embedded systems. Dr. Sethi has published numerous manuscripts in reputable journals, conferences and books. Currently, he is also actively involved in the technical program committees of various international conferences. He is serving as an associate editor for *EAI Endorsed Transactions on Context-Aware Systems and Applications* and at *EAI Endorsed Transactions on Ambient Systems.*

List of Abbreviations

ARQ	Automatic Retransmission Request
ASIC	Application-Specific Integrated Circuits
BE	Best Effort
CRC	Cyclic Redundancy Codes
DN	Direct Neighbor
DSP	Digital Signal Processing
FEC	Forward Error Correction
FPGA	Field Programmable Gate Arrays
GALS	Globally Asynchronous Locally Synchronous
GT	Guaranteed Throughput
HPC	High-Performance Computing
IN	Immediate Neighbor
IP	Internet Protocol
ISE	Information Set East
ISN	Information Set North
ISS	Information Set South
ISW	Information Set West
MPSoC	Multiprocessor System on Chip
NoC	Network on Chip
PE	Processing Element
QoS	Quality of Service
SoC	System on Chip
SQL	Structured Query Language
TDM	Time Division Multiplexing
VC	Virtual Channel

List of Symbols

bps	bits per second
MBps	megabyte per second
MW	megawatt
µm	micrometer
µs	microseconds
µW	microwatt
mm²	millimeters squared
mW	milliwatt
nm	nanometer
ns	nanoseconds
pj	petajoule
pj/packet	petajoule per packet
ps	picoseconds

1 Introduction

1.1 NETWORK ON CHIP

Network on Chip (NoC) is a communication standard for on-chip communication. NoC has replaced the crossbar interconnections and bus with a network of wires. Latency, congestion and delayed communication are the drawbacks of the bus. Due to the modular structure of the NoC, it can effectively reuse these resources. NoC also has other benefits of higher bandwidth, concurrency and scalability. The building blocks of the NoC are routers, processing elements (PEs), network interfaces (NI) and interconnects. Routers are connected with each other and with the PE (local core). The NI separates data communications from network communications. The message generated by the PE is converted into packets by the NI and at the destination; packets are again converted to messages by NI. The router receives the packets from NI and later routes the packets in the direction of the destination based on the rules of the routing algorithm (Benini and De Micheli 2002). Figure 1.1 shows a 4 × 4 mesh NoC with 16 routers and PEs. PEs can be homogeneous or heterogeneous resources. PEs can be a processor (P), memory (M), cache (C), reconfigurable block (re), digital signal processing (DSP) core or any other intellectual property cores, as shown in Figure 1.2.

In an NoC, regular and irregular are two major types of topologies through which PEs can be connected together. PEs and routers are connected in a systematic pattern in a regular topology. Moreover, an irregular topology has no systematic and structured pattern of PEs and routers. Figure 1.3 and Figure 1.4 show the NoC torus, tree, ring and irregular topologies.

The communication issues between multiple PEs are solved by NoC. However, as the number of devices is increasing on the chip, the NoC has encountered various communication issues in the form of faults. Devices are suffering from permanent and temporary faults due to the shrinking of device dimensions. Permanent faults can only be removed by redundant hardware, while transient faults can be overcome by routing algorithms. These routing algorithms are used to having reliable and efficient communication between multiple PEs by avoiding faulty routers and interconnects (Safaei and ValadBeigi 2012). In this research study, the focus is on recovering from temporary faults using fault-tolerant routing algorithms.

Fault tolerance is the basic concept, which separates the traditional data communication network from the on-chip communication network. The Internet Protocol (IP) is one of the complex algorithms that provides fault tolerance in data networks using cyclic redundancy codes (CRC), forward error correction (FEC) and automatic retransmission request (ARQ). IP requires a lot of resources and area, which are unavailable in the on-chip network. Moreover, the low latency requirement is another characteristic of NoC, which is not guaranteed by the IP network (Bogdan, Dumitraş, and Marculescu 2007).

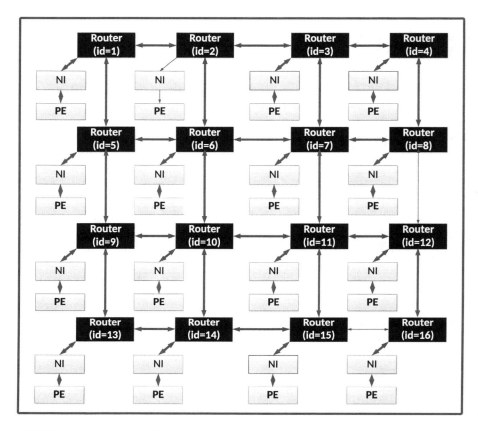

FIGURE 1.1 Network on Chip (NoC).

1.2 FAULT TOLERANCE

Fault tolerance is necessary for reliable communication between the source and destination. It is a vital issue, as devices on the chip are decreasing in size due to nanotechnology. Permanent faults are due to manufacturing defects, electromigration and dielectric breakdown or any other physical damage. Permanent faults can only be removed by extra hardware, that is, spare routers, wires or resources available on the NoC (Chang et al. 2011; Lehtonen et al. 2009; Koibuchi et al. 2008).

The temporary faults may occur due to changes in voltage, temperature fluctuations and congestion at the routers. Transient faults can be dealt with using fault-tolerant routing algorithms (Pasricha and Zou 2011; Nunez-Yanez, Edwards, and Coppola 2008; Kim and Kim 2007; Schonwald et al. 2007; Rantala, Lehtonen, and Plosila 2006; Zhu, Pande, and Grecu 2007; Pirretti et al. 2004; Wu 2000; Schonwald, Bringmann, and Rosenstiel 2007; Glass and Ni 1992; Chiu 2000; Patooghy and Miremadi 2010; Chien and Kim 1992). Fault-tolerant routing algorithms are used

Introduction

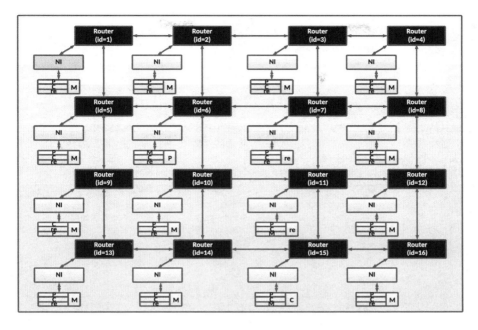

FIGURE 1.2 NoC with heterogeneous resources.

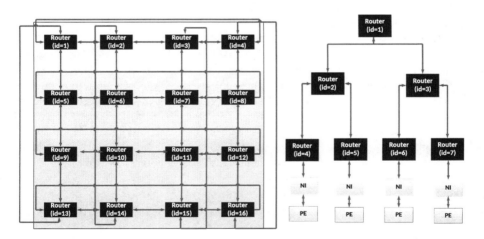

FIGURE 1.3 NoC torus and tree topologies.

to recover from the faulty interconnects in the NoC. These algorithms ensure reliable and efficient communication between the source PE and destination PE. Deterministic, stochastic, fully adaptive and partial adaptive routing algorithms are the four broad categories used.

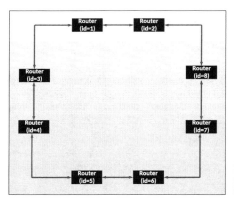 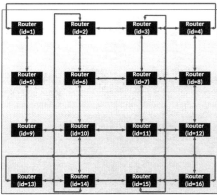

FIGURE 1.4 NoC ring and irregular topologies.

1.3 WHY USE BIO-INSPIRED ALGORITHMS FOR FAULT TOLERANCE

In deterministic routing, the packet routes from a certain point to another using a fixed path. These algorithms are simple to implement and deadlock-free; however, they are not fault-tolerant, lack adaptability and are not suitable for complex networks. In stochastic routing algorithms, a packet is sent either in all directions or in a particular direction based on the type of routing algorithm. These algorithms are simple to implement, but they consume a great deal of bandwidth and are not fault-tolerant. In full adaptive routing algorithms, the routing depends on a table situated at the router. These algorithms are fault-tolerant, adaptive and very dynamic in nature, but updating the routing information takes a lot of time (latency), which affects the throughput of the NoC and consumes a lot of bandwidth. Finally, partial adaptive routing algorithms place restrictions on particular turns in the NoC. These algorithms solve the problem of deadlock and consume less bandwidth, as there are no routing tables. These algorithms are fault-tolerant and less complicated but lack the adaptability, end-to-end latency and interflit arrival time of the packets increases due to restrictions on particular turns in the NoC. Details of different fault-tolerant algorithms in the literature are described in Section 2.1 and summarized in Table 1.1.

The majority of the traditional fault-tolerant algorithms do not entirely address the faults in the NoC and have the drawbacks of high end-to-end latency and inter-\ flit arrival time, low bandwidth utilization and less throughput. This makes the NoC communication unreliable, and the NoC architecture is not fault-tolerant. To overcome the drawbacks of deterministic, stochastic, fully adaptive and partially adaptive routing algorithms, novel biological inspired fault-tolerant algorithms were proposed. Bio-inspired algorithms can address the limitations of fault-tolerant algorithms by mimicking the fault-tolerant and robust mechanism of the brain based

Introduction

TABLE 1.1
NoC Literature on Fault-Tolerant Algorithms

Fault-Tolerant Routing Algorithms

Categories	Types	Pros	Cons
Deterministic routing algorithms	1. XY 2. YZ 3. XYZ 4. ZYX 5. Hybrid look-ahead fault-tolerant routing (HLAFT) along with random access buffer (RAB) deadlock recovery technique	• Simple and easy to implement. • Deadlock-free.	• Lacks adaptability. • Not fault-tolerant. (HLAFT has better routing choices and fault tolerance due to look-ahead and local routing mechanism.)
Stochastic routing algorithms	1. Probabilistic gossip flooding scheme. 2. Directed flooding scheme. 3. N-random walk. 4. Connection-oriented stochastic routing (COSR).	Simple and easy to implement.	• High power consumption. • Congestion. • Deadlocks and livelocks. • Consumes a lot of bandwidth. • Not dynamic in nature. • Not fault-tolerant.
Fully adaptive routing algorithms	1. Source routing for NoC (SRN). 2. Force-directed wormhole routing (FDWR). 3. Fault-tolerant deflection routing (FTDR) and hierarchical fault-tolerant deflection routing (FTDR-H). 4. Fault-tolerant-oASIS (FTO), the adaptive fault-tolerant routing algorithm. 5. Scalable fault-tolerant routing algorithm.	• Dynamic in nature. • Low-latency communication. • Flits communication possible. • Fault-tolerant.	• Continually updating the routing table. • Communication between routers. • Routing table requires a lot of areas. • Congestion at the router.
Partial adaptive routing algorithms	1. West first, negative first, north last, south last. 2. Odd-even. 3. Planar adaptive. 4. Hexagonal NoC partial adaptive routing.	• Less power consumption. • No deadlock and livelock. • No routing table. • Fault-tolerant.	• Lacks adaptability.

on the algorithms called "synaptogenesis" and "sprouting". These bio-inspired NoC fault-tolerant algorithms have the following characteristics:

1. Bio-inspired algorithms do not pursue any fixed way toward a destination in contrast to a deterministic routing algorithm.
2. The proposed bio-inspired algorithms do not broadcast the packet (flits) in any direction that efficiently utilizes the bandwidth compared to stochastic routing algorithms.
3. There is no routing table in these bio-inspired algorithms, yet they are highly adaptive and robust compared to fully adaptive algorithms.
4. There is no constraint on turns in the NoC. However, it takes two nodes' neighbor information before making a decision on the turn to take in contrast to partial adaptive algorithms.

1.4 OBJECTIVES OF BIO-INSPIRED FAULT-TOLERANT ALGORITHMS

1. To develop the bio-inspired NoC fault-tolerant algorithm(s) to monitor the routers and interconnects between PEs (cores) to address the limitation in the literature on fault-tolerant algorithms. The bio-inspired fault-tolerant algorithms can:
   - Self-adapt (using synaptogenesis)
   - Self-heal (using sprouting)
   - Self-optimize (using best-effort and guaranteed throughput services)
2. To design the NoC architecture that adopt bio-inspired algorithm(s) to make the NoC architecture fault-tolerant and reliable.

1.5 BOOK CONTRIBUTIONS

1. Implementation of a fault-tolerant synaptogenesis algorithm using best-effort services. In this algorithm, packet switching was used to traverse the flits over the NoC. The fault was detected using credit-based flow control.
2. Implementation of a fault-tolerant sprouting algorithm using best-effort services. In this algorithm, packet switching was used to traverse the flits over NoC. The fault detection mechanism is improved by the sprouting. The fault was detected by the neighboring router when a flit was received on the blocked port.
3. Implementation of a bio-inspired fault-tolerant algorithm using the guaranteed throughput approach. In this algorithm, a connection-oriented mechanism was used for traversing flits over the NoC. The fault was detected by the neighboring router as the flit was received on the blocked port. No flow control technique was required, as the flow was synchronized by a clock.
4. Implementation of a bio-inspired fault-tolerant algorithm using combined best-effort and guaranteed throughput services. In this algorithm, the connectionless and connection-oriented mechanisms of packet switching were combined. The fault was detected by a neighboring router as a flit was

Introduction

received on the blocked port. For guaranteed throughput, no flow control technique was required, while the credit-based flow control technique was needed for the connectionless mechanism.

5. Review of the NoC architectures based on different parameters.

1.6 ORGANIZATION OF THE BOOK

Chapter 2 of this book provides a literature review on NoC fault-tolerant routing algorithms. Moreover, this chapter also explains the different parameters of the NoC used in the bio-inspired fault-tolerant algorithms.

Chapter 3 is about biologically inspired algorithms solving real-world problems. Moreover, this chapter also discusses the implementation of biological mechanisms in the NoC to address various issues. Bio-inspired fault-tolerant routing algorithms for NoC; bio-inspired NoC framework; bio-inspired NoC network; and bio-inspired NoC algorithms using BE, GT and combined BE-GT are described in Chapter 4.

Chapter 5 describes the experimental results. In this chapter, the bio-inspired algorithm, including the literature algorithms, are thoroughly analyzed and compared and different conclusions are derived.

The book is concluded in Chapter 6. In this chapter, the overall summary of the book is presented along with future recommendations. The specific details of survey studies related to each of NoC architectures are provided in Appendix A, while Appendix B contains the actual fault detection code.

REFERENCES

Atienza, David, Federico Angiolini, Srinivasan Murali, Antonio Pullini, Luca Benini, and Giovanni De Micheli. 2008. "Network-on-chip design and synthesis outlook." *Integration* 41 (3):340–359.

Benini, Luca, and Giovanni De Micheli. 2002. "Networks on chips: A new SoC paradigm." *Computer* 35 (1):70–78.

Bogdan, Paul, Tudor Dumitraș, and Radu Marculescu. 2007. "Stochastic communication: A new paradigm for fault-tolerant networks-on-chip." *VLSI Design 2007*.

Chang, Yung-Chang, Ching-Te Chiu, Shih-Yin Lin, and Chung-Kai Liu. 2011. "On the design and analysis of fault tolerant NoC architecture using spare routers." *Proceedings of the 16th Asia and South Pacific Design Automation Conference*.

Chien, Andrew A, and Jae H Kim. 1992. *Planar-adaptive routing: Low-cost adaptive networks for multiprocessors*. Vol. 20: ACM.

Chiu, Ge-Ming. 2000. "The odd-even turn model for adaptive routing." *IEEE Transactions on Parallel and Distributed Systems* 11 (7):729–738.

Glass, Christopher J, and Lionel M Ni. 1992. "The turn model for adaptive routing." *ACM SIGARCH Computer Architecture News* 20 (2):278–287.

Guerrier, Pierre, and Alain Greiner. 2008. "A generic architecture for on-chip packet-switched interconnections." *Design, Automation, and Test in Europe*.

Kim, Young Bok, and Yong-Bin Kim. 2007. "Fault tolerant source routing for network-on-chip." *22nd IEEE International Symposium on Defect and Fault-Tolerance in VLSI Systems (DFT 2007)*.

Koibuchi, Michihiro, Hiroki Matsutani, Hideharu Amano, and Timothy Mark Pinkston. 2008. "A lightweight fault-tolerant mechanism for network-on-chip." *Proceedings of the Second ACM/IEEE International Symposium on Networks-on-Chip*.

Lehtonen, Teijo, David Wolpert, Pasi Liljeberg, Juha Plosila, and Paul Ampadu. 2009. "Self-adaptive system for addressing permanent errors in on-chip interconnects." *IEEE Transactions on Very Large Scale Integration (VLSI) Systems* 18 (4):527–540.

Nunez-Yanez, José L, Doug Edwards, and Antonio Marcello Coppola. 2008. "Adaptive routing strategies for fault-tolerant on-chip networks in dynamically reconfigurable systems." *IET Computers & Digital Techniques* 2 (3):184–198.

Pasricha, Sudeep, and Yong Zou. 2011. "A low overhead fault tolerant routing scheme for 3D Networks-on-Chip." *2011 12th International Symposium on Quality Electronic Design*.

Patooghy, Ahmad, and Seyed Ghassem Miremadi. 2010. "Complement routing: A methodology to design reliable routing algorithm for network on chips." *Microprocessors and Microsystems* 34 (6):163–173.

Pirretti, Matthew, Greg M Link, Richard R Brooks, Narayanan Vijaykrishnan, Mahmut Kandemir, and Mary Jane Irwin. 2004. "Fault tolerant algorithms for network-on-chip interconnect." *IEEE Computer Society Annual Symposium on VLSI*.

Rantala, Ville, Teijo Lehtonen, and Juha Plosila. 2006. *Network on chip routing algorithms*: Citeseer.

Safaei, Farshad, and Majed ValadBeigi. 2012. "An efficient routing methodology to tolerate static and dynamic faults in 2-D mesh networks-on-chip." *Microprocessors and Microsystems* 36 (7):531–542.

Schonwald, Timo, Oliver Bringmann, and Wolfgang Rosenstiel. 2007. "Region-based routing algorithm for network-on-chip architectures." *Norchip 2007*.

Schonwald, Timo, Jochen Zimmermann, Oliver Bringmann, and Wolfgang Rosenstiel. 2007. "Fully adaptive fault-tolerant routing algorithm for network-on-chip architectures." *10th EUROMICRO Conference on Digital System Design Architectures, Methods and Tools (DSD 2007)*.

Seifi, Mohammad Reza, and Mohammad Eshghi. 2012. "Clustered NoC, a suitable design for group communications in Network on Chip." *Computers & Electrical Engineering* 38 (1):82–95.

Wu, Jie. 2000. "A fault-tolerant adaptive and minimal routing approach in nD meshes." *Proceedings 2000 International Conference on Parallel Processing*.

Zhu, Haibo, Partha Pratim Pande, and Cristian Grecu. 2007. "Performance evaluation of adaptive routing algorithms for achieving fault tolerance in NoC fabrics." *2007 IEEE International Conf. on Application-specific Systems, Architectures and Processors (ASAP)*.

2 In-Depth Review of Network on Chip

2.1 LINK-SHARING MECHANISM

The link-sharing mechanism defines the way a communication channel is shared between multiple sources and destinations in the network on chip (NoC). Link-sharing mechanisms are divided into two broad categories: circuit switching and packet switching (connection-oriented and connectionless). According to the literature, most of the NoC architectures use packet switching, whereas only a few use circuit switching. This clearly shows the interest in the effective and efficient utilization of resources compared to allocating the resources for particular time periods. In circuit switching–based connections, the resources can be underutilized at certain times. Two NoC architectures use spatial division multiplexing (SDM)–based packet-switching techniques, while another uses wavelength division multiplexing (WDM) for link sharing.

The circuit-switching connections also lack the ability to react rapidly to specific changes in bandwidth and throughput requirements by processing elements (PEs) (Guerrier and Greiner 2008). To efficiently utilize the bandwidth of the NoC, the concept of virtual channels has been introduced using time division multiplexing (Moraes et al. 2004), frequency division multiplexing or a combination of both techniques (Ganguly et al. 2010). Despite the drawbacks of the circuit-switching technique, it is still being used by NoC applications that require guaranteed services (Ogras, Hu, and Marculescu 2005). Future link-sharing mechanisms should include the adaptive and hybridized mechanisms, which can switch between circuit and packet switching based on the traffic load and PE requirements.

2.1.1 CIRCUIT SWITCHING

In circuit switching, a physical link is established between the source and destination before data transmission. After the connection is established, flits (flow control digits) or packets traverse various routers. The connection is established until all packets are received at the destination (Nurmi 2005). The circuit-switching technique is better for high-traffic, real-time applications. These applications generate and send traffic at a higher injection rate. As there is a separate link between the source and destination, there is no communication delay problem and high throughput can be achieved. In circuit switching, the bandwidth is reserved for the entire duration of data transmission, but resources (routers and links) are busy until all the data are received at the destination. The setup of the path from source to destination increases the unnecessary delay (Pande et al. 2005). NoC architectures which are based on circuit-switching techniques include (Lines 2004; Hilton and Nelson 2006; Chang,

Shen, and Chen 2006; Dally and Towles 2001; Henriksson, Wiklund, and Liu 2003; Peña-Ramos and Parra-Michel 2011; Wiklund and Liu 2003; Wolkotte et al. 2005; Butts 2007; Ainsworth and Pinkston 2007; Braun et al. 2007). The Dally et al. (Dally and Towles 2001) NoC architecture provides virtual channel–based circuit switching while OCN (Henriksson, Wiklund, and Liu 2003) provides circuit switching using round-robin arbitration. The only drawback of the techniques that adopt circuit switching is the average utilization of the link. The links are underutilized at certain times. Therefore, to address these issues, packet switching must also be used in these systems to efficiently utilize the communication channel (Rijpkema et al. 2003).

2.1.2 PACKET SWITCHING

In packet switching, routers decide in which direction of the NoC a packet should be sent based on the routing algorithm. In packet switching, the messages generated by the source are divided into packets and flits. The reason for dividing the messages into packets and flits is to utilize the bandwidth of the NoC efficiently. The flits can traverse the multiple paths available on the NoC (Wang et al. 2011). Several NoC architectures are based on packet switching: (Guerrier and Greiner 2008; Moraes et al. 2004; Lines 2004; Bolotin et al. 2004; Wang, Ahonen, and Nurmi 2007; Bainbridge and Furber 2002; Siguenza-Tortosa and Nurmi 2002; Kumar et al. 2002; Taylor et al. 2002; Zeferino and Susin 2003; Samuelsson and Kumar 2004; Zeferino, Kreutz, and Susin 2004; Chan and Parameswaran 2004; Bartic et al. 2005; Neeb and Wehn 2008; Kariniemi and Nurmi 2006; Mullins, West, and Moore 2006; Zid et al. 2006; Bell et al. 2008; Hoffman et al. 2007; Janarthanan, Swaminathan, and Tomko 2007; Hosseinabady et al. 2007; Soteriou et al. 2006; Kumar et al. 2008; Samman, Hollstein, and Glesner 2009; Feero and Pande 2008; Lan et al. 2011; Krasteva, De la Torre, and Riesgo 2010; Wu, Tang, and Hsu 2011; Mishra, Nidhi, and Kishore 2012; Shu et al. 2012; Choudhary and Qureshi 2012; Arteris 2005; Coppola et al. 2004; Feliciian and Furber 2004; Beigné et al. 2005; Panades, Greiner, and Sheibanyrad 2006; Schuck, Lamparth, and Becker 2007; Beigné et al. 2009; Forsell 2002; Ching, Schaumont, and Verbauwhede 2004; Hu and Marculescu 2004; Anjo et al. 2004; Lee, Lee, and Lee 2004; Bobda et al. 2005; Sanusi and Wang 2006; Pionteck, Koch, and Albrecht 2006; Ogras et al. 2006; Palermo et al. 2007; Lattard et al. 2008; Wang et al. 2010; Lee et al. 2012; Kao and Chao 2012; Yang et al. 2014; Al Faruque, Ebi, and Henkel 2010). These packet-switching NoC architectures either use round-robin arbitration, virtual channels or any other mechanisms, and specific details can be found in Appendix A. Figure 2.1 shows that multiple parallel virtual channels are created between various source PEs and destination PEs through a connection-oriented mechanism.

SDM and WDM are the link-sharing techniques used by some NoC architectures. SDM physically divides every link and buffer into multiple virtual circuits. Every virtual circuit is assigned a portion of the bandwidth. SDM uses wormhole switching for flow control (Song and Edwards 2011). An architecture and compiler for aSoC (Liang et al. 2004) and SDM NoC (Leroy et al. 2005) are two examples of SDM. WDM can send multiple signals simultaneously to attain higher throughput. The wavelength determines the destination address in WDM or routing. This makes

In-Depth Review of Network on Chip

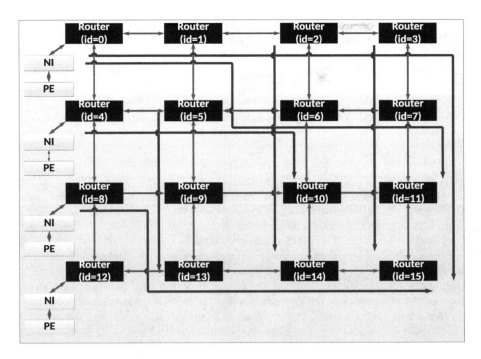

FIGURE 2.1 Multiple virtual channels (some PEs and network interfaces [NIs] are removed for clarity).

the WDM the contention-free link-sharing mechanism. The ORNoC (Le Beux et al. 2011), NoC architecture provides link sharing using WDM. Some of the architectures have not provided any information regarding their link-sharing mechanism. The details of the architectures are provided in Appendix A.

2.2 NoC FAULT-TOLERANT ROUTING ALGORITHMS

The routing algorithm defines the path for a packet between the source PE and destination PE. Fault-tolerant algorithms are used to recover from a faulty interconnect in the NoC by ensuring reliable and efficient communication between the source PE and destination PE. There are basically four broad categories of routing algorithms. They are deterministic, stochastic, fully adaptive and partial adaptive routing algorithms. When the routing decision is made by the source, it is called source routing, whereas if the immediate nodes of the NoC make the decision, then the routing algorithm is categorized as distributed routing (Moraes et al. 2004).

Deterministic routing algorithms deliver an orderly packet and require fewer resources compared to adaptive routing algorithms. Adaptive routing algorithms provide better throughput and low latency by having alternative paths due to congested or faulty routes. Deterministic and partially adaptive algorithms are deadlock and livelock free, whereas fully adaptive algorithms require some precaution by having deadlock-, livelock- and congestion-avoiding techniques (Ogras, Hu, and Marculescu

2005). Adaptive routing algorithms need specialized modules at the receiver to reorder the packets, which in turn increases the design complexity and latency of the packets. Deterministic routing algorithms perform well under uniform traffic patterns, while adaptive routing algorithms are preferred for bursty and irregular traffic (Palesi and Daneshtalab 2014). A small number of NoC architectures have not specified nor is it clear which routing algorithm they are using. In the literature, deterministic routing algorithms have been used because they are less complex to implement.

2.2.1 DETERMINISTIC ROUTING ALGORITHMS

In deterministic routing the packet routes from a certain point to another using a fixed path. These algorithms lack adaptivity, and they are not fault-tolerant. XY, YX, XYZ and ZYX (Pasricha and Zou 2011; Kim and Kim 2007; Schonwald et al. 2007; Zhu, Pande, and Grecu 2007; Schonwald, Bringmann, and Rosenstiel 2007; Glass and Ni 1992; Chiu 2000) are a few examples of dimension order routing (DOR) algorithms. DOR algorithms are the simplest deterministic routing algorithms, and they are deadlock-free. In minimal path routing, the packet can traverse using multiple (shortest) paths to reach the destination. Minimum path routing algorithms are prone to deadlocks compared to DOR (Bahrebar and Stroobandt 2014).

In the XY routing algorithm, the packet is first reached at the row of the destination node and then to the column of the destination node. In YX, the packet is first routed to the column of the destination node and then to the row. These algorithms are static, deadlock-free and deterministic. They lack adaptivity and are not suitable for sophisticated network resources. ZYX is the 3D NoC routing algorithm, and in this algorithm, the packet is first sent to the NoC layer, then it is routed to the column of the destination and later it is sent to the particular row of the destination (Pasricha and Zou 2011). The ZYX algorithm is static, deterministic and deadlock-free, but it lacks adaptivity. Hybrid look-ahead fault-tolerant routing (HLAFT) is a deterministic routing algorithm as it utilizes both local and look-ahead routing to ensure fault tolerance in the NoC. HLAFT also uses a random access buffer (RAB) mechanism to avoid deadlock in the NoC. HLAFTs have lower latency and higher throughput compared to DOR (Ahmed and Abdallah 2014).

NoC architectures that have deterministic routing are (Moraes et al. 2004; Liang, Swaminathan, and Tessier 2000; Karim, Nguyen, and Dey 2002; Lines 2004; Bolotin et al. 2004; Leroy et al. 2005; Hilton and Nelson 2006; Chang, Shen, and Chen 2006; Salminen et al. 2006; Castells-Rufas, Joven, and Carrabina 2006; Janarthanan and Tomko 2008; Dally and Towles 2001; Kumar et al. 2002; Henriksson, Wiklund, and Liu 2003; Zeferino and Susin 2003; Bertozzi and Benini 2004; Zeferino, Kreutz, and Susin 2004; Bartic et al. 2005; Kariniemi and Nurmi 2006; Bell et al. 2008; Gratz et al. 2007; Hoffman et al. 2007; Janarthanan, Swaminathan, and Tomko 2007; Soteriou et al. 2006; Kumar et al. 2008; Samman, Hollstein, and Glesner 2009; Lan et al. 2011; Krasteva, De la Torre, and Riesgo 2010; Peña-Ramos and Parra-Michel 2011; Shu et al. 2012; Al Faruque, Ebi, and Henkel 2010; Choudhary and Qureshi 2012; Evain, Diguet, and Houzet 2004; Coppola et al. 2004; Bjerregaard and Sparso 2004, 2005a, 2005c; Beigné et al. 2005; Panades, Greiner, and Sheibanyrad 2006; Schuck, Lamparth, and Becker 2007; Wang and Bagherzadeh 2014; Heisswolf et al.

In-Depth Review of Network on Chip

2015; Forsell 2002; Hu and Marculescu 2004; Chi and Chen 2004; Bobda et al. 2005; Sanusi and Wang 2006; Pionteck, Koch, and Albrecht 2006; Wang et al. 2010; Bahirat and Pasricha 2014; DiTomaso, Kodi, and Louri 2014; Poluri and Louri 2014).

2.2.2 STOCHASTIC ROUTING ALGORITHMS

Packets are broadcast in either all directions or a particular direction of the NoC depending on the type of algorithm. These algorithms are simple to implement, but they consume a lot of power. They have congestion, deadlock, livelock and high-bandwidth utilization problems. These algorithms do not perform well even at low traffic rates because they broadcast the packet in all directions. These algorithms are not fault-tolerant, lack adaptivity and are not dynamic in nature. A few examples of stochastic algorithms are probabilistic gossip flooding scheme, directed flooding, N-random walk (Pasricha and Zou 2011; Zhu, Pande, and Grecu 2007; Pirretti et al. 2004; Patooghy and Miremadi 2010) and connection-oriented stochastic routing (COSR) (Nunez-Yanez, Edwards, and Coppola 2008). None of the NoC architectures has used the stochastic routing algorithm due to its drawbacks.

2.2.2.1 Probabilistic Gossip Flooding Scheme

In this algorithm, packets are sent multiple times to multiple paths of the NoC. These algorithms do not have any information about the destination. When one of the copies of the packet arrives at the destination, all other packet copies are deleted. This algorithm consumes a lot of bandwidth and power, lacks adaptivity and is not dynamic in nature. Also, the packets in this scheme may get lost or collapse during the traversal of the NoC (Zhu, Pande, and Grecu 2007; Pirretti et al. 2004).

2.2.2.2 Directed Flooding Scheme

Packets are replicated in particular or directed paths in the directed flooding scheme. In terms of performance, this algorithm is better than the probabilistic algorithm, as it consumes a small amount of bandwidth and few network resources (Pasricha and Zou 2011). This algorithm is less influenced by higher error rates. The performance is better than probabilistic algorithms when the gossip rate is low (Kim and Kim 2007).

2.2.2.3 N-random Walk

(N) number of packets are sent to the network in a particular direction in the N-random walk algorithm. This algorithm is efficient compared to directed flooding, as it consumes a small amount of bandwidth and power (Zhu, Pande, and Grecu 2007). The random walk has less communication overhead compared to the flooding scheme, and it also provides a useful level of fault tolerance (Bjerregaard and Mahadevan 2006).

2.2.2.4 Connection-Oriented Stochastic Routing

This scheme is a hybrid algorithm that combines circuit and packet switching. A connection is established using circuit switching between the source and destination. After a connection is established, packets are sent over the established connection using packet switching (Nunez-Yanez, Edwards, and Coppola 2008). This algorithm is easy to implement and is fault-tolerant. The drawback of this algorithm is that the connection established for a certain period of time may lead to underutilization of the bandwidth, congestion and denial of services to some PEs.

2.2.3 Fully Adaptive Routing Algorithms

The routing of fully adaptive algorithms depends upon the routing table or on the routing information collected from the neighbor nodes at the router. Based on this information, the direction of the packet is decided at run time. Routers constantly communicate with each other to update the routing table or neighbor node information. The updating of routing information takes a lot of power and time, consequently affecting the throughput of the NoC (Pasricha and Zou 2011; Rantala, Lehtonen, and Plosila 2006). Source routing for NoC (SRN) and force-directed wormhole routing (FDWR) are two examples of fully adaptive routing algorithms (Kim and Kim 2007; Schonwald et al. 2007). Fully adaptive routing algorithms are fault-tolerant and very dynamic, but the updating of routing information consumes a lot of area and power, which sometimes degrade the performance of the NoC. The flow of control packets between routers at times creates congestion and deadlock situations in the NoC. The routing algorithms are fully adaptive, but they do not have a routing table. They collect the neighbor's information through certain control packets. Based on this information, the router decides in which direction the packet should be routed.

A few NoC architectures using source-based routing are (Guerrier and Greiner 2008; Rijpkema et al. 2003; Paukovits and Kopetz 2008; Hansson, Subburaman, and Goossens 2009; Bainbridge and Furber 2002; Siguenza-Tortosa and Nurmi 2002; Pande et al. 2003; Kim et al. 2005; Kariniemi and Nurmi 2006; Zid et al. 2006; Vangal et al. 2008; Hosseinabady et al. 2007; Soteriou et al. 2006; El-Moursy, Korzec, and Ismail 2009; Krasteva, De la Torre, and Riesgo 2010; Yaghini, Eghbal, and Bagherzadeh 2015; Bjerregaard and Sparso 2004, 2005a, 2005c; Kavaldjiev, Smit, Jansen et al. 2006; Mondinelli, Borgatti, and KOVACS VAJNA 2004; Lee et al. 2005; Bouhraoua and Elrabaa 2006; Ahmad, Erdogan, and Khawam 2006).

2.2.3.1 Source Routing for NoC

Router discovery and route maintenance are two main components of this algorithm. First, the algorithm dynamically discovers its path from the source to destination. Later, it maintains the path by adapting another new path from the source to bypass the faulty router. Due to a limited number of broadcasts, SRN has less communication overhead and is not able to handle complex networks. It has a restriction on its source routing table, which limits its adaptivity (Kim and Kim 2007).

2.2.3.2 Force-Directed Wormhole Routing

FDWR divides the traffic over the network by force. It follows the wormhole-switching mechanism. In wormhole switching, the packet is divided into header, body and tail flits. The header flit finds and contains the path until the destination. The body and tail flit follow the header flit's specified path. Body flits contain the data to be transferred from the source to destination, while tail flits inform the routers and the

In-Depth Review of Network on Chip   **15**

destination about the termination of communication. In FDWR, the header flit finds the path by avoiding the faulty routers. The neighbor routers, which are working correctly, exchange messages about their availability and working status. If the routers receive the messages with the help of a routing table, FDWR finds the shortest path from the source to destination (Schonwald et al. 2007). The advantage of FDWR is that it evenly distributes the traffic across the network with the help of a routing table. It is fault-tolerant, as it avoids the faulty routers by taking another path. In this algorithm, the flits of the packets may choose different ways on the NoC. With the help of a sequence number, flits and packets are reordered at the destination. The drawback of this algorithm is that the destination has to reorder a packet, which consumes power and time.

Fault-tolerant deflection routing (FTDR) and the hierarchical-routing-table-based FTDR algorithm (FTDR-H) detects transients and permanent faults for bufferless NoCs across links. FTDR and FTDR-H results are better than stochastic routing algorithms. The drawback of these techniques is that when a fault is detected, the link is shut down in a bidirectional way, and transient faults are only valid for one cycle (Feng et al. 2012). Fault-tolerant-OASIS (FTO) detects and works around transient, intermittent and permanent faults in the NoC. FTO ensures fault tolerance in the NoC by using the least amount of additional hardware complexity. FTO performs better than DOR (Ahmed and Abdallah 2016). The scalable fault-tolerant routing algorithm detects permanent failures in the NoC links and reconfigures the routing path using active routers and links. The drawback of the routing algorithm is its complexity, but it provides multiple routing paths in case of faults (Kia et al. 2015).

The advantages of the fully adaptive routing algorithms are that they are dynamic in nature. The router decides at run time to which link it should route the packet. However, continually updating the routing table requires a lot of time, area and power. Communication between routers in a fully adaptive routing algorithm creates deadlock and congestion in the NoC (Pasricha and Zou 2011; Patooghy and Miremadi 2010).

2.2.4 PARTIAL ADAPTIVE ROUTING ALGORITHMS

Partial adaptive algorithms, as the name implies, are partially adaptive. They put some restrictions on the routes that can be taken by a router in the NoC (Moraes et al. 2004). These algorithms solve the problem of deadlock and consume less power, as there are no routing tables. These algorithms are fault-tolerant but limit the adaptivity of the NoC and the latency of the packets increases due to restrictions on turns (Pasricha and Zou 2011). West first, negative first, north last, south last, odd-even and planar adaptive (Pasricha and Zou 2011; Rantala, Lehtonen, and Plosila 2006; Zhu, Pande, and Grecu 2007; Wu 2000; Schonwald, Bringmann, and Rosenstiel 2007; Glass and Ni 1992; Chiu 2000; Chien and Kim 1992) are a few examples of these algorithms.

These NoC architectures (Janarthanan, Swaminathan, and Tomko 2007; Wu, Tang, and Hsu 2011; Yang et al. 2014; Choudhary and Qureshi 2012; Postman et al. 2012; Schuck, Lamparth, and Becker 2007) have turn-based adaptive routing algorithms. Odd-even routing algorithms are used by these (Taylor et al. 2002; Chan and

Parameswaran 2004; Hoffman et al. 2007; Samman, Hollstein, and Glesner 2009; Lan et al. 2011; Choudhary and Qureshi 2012; Evain, Diguet, and Houzet 2004; Beigné et al. 2005; Hu and Marculescu 2004; Lattard et al. 2008) NoC architectures. A few of the NoC architectures that have distributed routing algorithms are (Stefan, Molnos, and Goossens 2012; Samuelsson and Kumar 2004; Wiklund and Liu 2003; Wolkotte et al. 2005; Schoeberl 2007), while Nostrum (Penolazzi and Jantsch 2006; Millberg et al. 2004) has TDM-based deflective routing. Other NoC architectures follow stream-based routing, e-cube routing, Dijsktra shortest path routing, shortest distance, shortest path first and temperature first routing algorithms. The details of these and other NoC architectures are presented in Appendix A.

2.2.4.1 West First, Negative First, North Last and South Last Routing Algorithms

In the west first algorithm, packets transmitted to the west should be first because later in this algorithm, the packet cannot be sent in a westerly direction. Similarly, the negative first algorithm allows all other turns except turns from a positive direction to the negative direction. Packet routing in the negative direction must be done first, and all other turns in the NoC later. In the north last algorithm, packets that need to be routed north must be transferred last. However, in the south last algorithm, packets can be routed in a southerly direction as the last turn (Rantala, Lehtonen, and Plosila 2006).

2.2.4.2 Odd-Even Routing Algorithm

In the odd-even routing algorithm, no deadlock occurs. This algorithm imposes a restriction that at the even column, there should be no turns from east to north and from east to south. Similarly, in odd columns, there should be no turn from north to west and south to west. The advantage of this algorithm is that it avoids livelocks and consumes less power due to the restrictions on turns (Rantala, Lehtonen, and Plosila 2006; Chiu 2000).

One of the advantages of turn-based models is that no routing table is required. The router has to determine according to which turn algorithm it is routing the packets. Routers can send the packet in the wrong direction. In turn-based algorithms, routers become more complex. Among all partial adaptive routing algorithms, the odd-even routing algorithm provides more adaptivity, as there are fewer restricted turns.

2.2.4.3 Planar Adaptive Routing Algorithm

The essential requirement of a planar adaptive routing algorithm is that it has a constant number of virtual channels, and it does not depend on the size and dimension of the network (Wu 2000). This algorithm routes the packet in two dimensions until it reaches the destination. The algorithm avoids deadlock while also consuming less routing time because there is no routing table. The planar adaptive algorithm can be applied to (n) dimensional planes by allocating three virtual channels (Chien and Kim 1992). The disadvantage of this algorithm is that it requires extra virtual channels for communication. The hexagonal NoC partial adaptive routing algorithm (Moriam and Fettweis 2016) uses a turn model for routing, and it avoids

In-Depth Review of Network on Chip

the use of virtual channels. The algorithm is deadlock and livelock free, but does have increased complexity.

2.2.5 NoC Fault-Tolerant Architectures

The trends in using routing algorithms show that most of NoC architectures have used deterministic routing algorithms, while few NoC architectures are using fully and partial adaptive routing algorithms. However, none of the NoC architectures have used the stochastic routing algorithm. Less complex algorithms have been implemented, so that area and power consumption is less.

Dally et al. (Dally and Towles 2001) cover the transient faults that may occur due to errors in the packets. The transient faults can be recovered by checksum codes in the packet. PROTEO (Siguenza-Tortosa and Nurmi 2002) and Nexus (Lines 2004) claimed to be fault-tolerant, but no information is provided in the literature regarding it. Xpipes (Bertozzi and Benini 2004) also detects transient faults using distributed error detection bits. XGFT (Kariniemi and Nurmi 2006) detects the static, dynamic and transient faults using a fault diagnosis and repair (FDAR) algorithm. Other NoC architectures did not provide any information regarding fault tolerance in their architecture. Specific details can be found in Appendix A.

2.3 SWITCHING TECHNIQUES

The term "switching technique" refers to the control of messages (packets or flits) flowing between routers in the NoC. It helps the routing algorithm avoid congestion and conflicts between routers. Circuit-switching techniques do not require a forwarding strategy, as the resources are already reserved for them. Packet switching requires a forwarding strategy, as it has to decide on a per-node basis and it requires buffering. In packet switching, flits are saved in the router before any routing decision. There are broadly three types of switching techniques: store and forward, virtual cut-through and wormhole switching.

The buffering requirement of the virtual cut-through and store and forward techniques are one packet compared to a few flits in wormhole switching. The design complexity of virtual cut-through is high compared to the wormhole switching and store and forward technique. The cost (power consumption and area) of wormhole switching is lower compared to the virtual cut-through and store and forward technique (Tatas et al. 2014). At low traffic rates, virtual cut-through has the same low latency as wormhole switching, whereas at high traffic loads, the virtual cut-through has a high throughput similar to the store and forward technique (Bartic et al. 2005). It is a trade-off between buffering, design complexity and cost when using these switching techniques.

2.3.1 Store and Forward

In store and forward, the packet is completely received at the router and then a routing decision is made on it. The drawback of this technique is more storage requirements at the router. The latency of the packet also increases, as complete packets are stored before routing (Rantala, Lehtonen, and Plosila 2006). Few of the NoC architectures (Castells-Rufas, Joven, and Carrabina 2006; Samuelsson and Kumar 2004; Forsell 2002; Mondinelli, Borgatti, and KOVACS VAJNA 2004; Anjo et al. 2004) use store and forward flow control techniques.

2.3.2 VIRTUAL CUT-THROUGH

In virtual cut-through flow control, the flit is sent to the next router when the neighbor router guarantees that the complete packet can be saved. If there is no space in the neighbor router, then the entire packet will be kept in that particular router. Virtual cut-through routers also need buffers to save the whole packet, but they have less latency compared to the store and forward technique (Rantala, Lehtonen, and Plosila 2006; Moraes et al. 2004; Tatas et al. 2014). When no packets are blocked in the buffers, then virtual cut-through switching achieves the same latency as wormhole switching in NoC (Tatas et al. 2014). NoC architectures that have virtual cut-through as a flow control technique are (Castells-Rufas, Joven, and Carrabina 2006; Janarthanan and Tomko 2008; Bartic et al. 2005; Janarthanan, Swaminathan, and Tomko 2007; Grot et al. 2011; Mishra, Nidhi, and Kishore 2012; Lee et al. 2003; Pionteck, Koch, and Albrecht 2006; Qouneh et al. 2012).

2.3.3 WORMHOLE SWITCHING

In the wormhole-switching technique, a packet is divided into smaller pieces called flits (flow control digits). There are three types of flits; header, body and tail. The header flit finds and contains the routing information, whereas the body flit includes the data information. The body flits carry these flits from source to destination. The tail flit informs the routers about communication termination. The tail flit is received at the destination, which tells it about communication completion from a particular source. Wormhole switching has the least buffering requirement compared to the other two flow control techniques, as the router sends the flit to neighbor routers even if it has a space for a single flit. Wormhole switching has the least latency of packets compared to other techniques, as flits may follow multiple paths on the NoC (Rantala, Lehtonen, and Plosila 2006; Rijpkema et al. 2003). In some instances, the routing and control information is also provided in the packet by allocating a few control bits (Bertozzi and Benini 2004). The drawback of this technique is that when the header flit is blocked in any router, all the subsequent flits of a packet in multiple routers are also blocked. This leads to a deadlock situation in the NoC (Tatas et al. 2014). Few architectures have separate control lines, and others use the flow of tokens for flow control between routers (Benini and De Micheli 2002; Kao and Chao 2012).

NoC architectures that have adopted wormhole switching as a flow control technique are (Guerrier and Greiner 2008; Moraes et al. 2004; Rijpkema et al. 2003; Karim, Nguyen, and Dey 2002; Bolotin et al. 2004; Castells-Rufas, Joven, and Carrabina 2006; Wang, Ahonen, and Nurmi 2007; Dally and Towles 2001; Kumar et al. 2002; Taylor et al. 2002; Pande et al. 2003; Zeferino and Susin 2003; Bertozzi and Benini 2004; Zeferino, Kreutz, and Susin 2004; Chan and Parameswaran 2004; Kim et al. 2005; Neeb and Wehn 2008; Mullins, West, and Moore 2006; Zid et al.

In-Depth Review of Network on Chip

2006; Bell et al. 2008; Gratz et al. 2007; Vangal et al. 2008; Hoffman et al. 2007; Hosseinabady et al. 2007; Soteriou et al. 2006; El-Moursy, Korzec, and Ismail 2009; Samman, Hollstein, and Glesner 2009; Feero and Pande 2008; Lan et al. 2011; Krasteva, De la Torre, and Riesgo 2010; Wu, Tang, and Hsu 2011; Shu et al. 2012; Yang et al. 2014; Al Faruque, Ebi, and Henkel 2010; Choudhary and Qureshi 2012; Arteris 2005; Yaghini, Eghbal, and Bagherzadeh 2015; Coppola et al. 2004; Feliciian and Furber 2004; Beigné et al. 2005; Panades, Greiner, and Sheibanyrad 2006; Schuck, Lamparth, and Becker 2007; Wang and Bagherzadeh 2014; Heisswolf et al. 2015; Ching, Schaumont, and Verbauwhede 2004; Hu and Marculescu 2004; Bouhraoua and Elrabaa 2006; Lee et al. 2006; Ahmad, Erdogan, and Khawam 2006; Ogras et al. 2006; Ainsworth and Pinkston 2007; Lattard et al. 2008; Lee et al. 2012).

Wolkotte et al. (Wolkotte et al. 2005) has separate control lines for an end-to-end flow control, while some architectures use the flow of tokens for flow control between routers (Postman et al. 2012; Wolkotte et al. 2005). The SWIFT (Postman et al. 2012) NoC architecture has token-based flow control.

2.4 BUFFER MANAGEMENT TECHNIQUES

Buffer management techniques are an important component in NoC to have reliable communication between sources and destinations. These techniques help to avoid buffer overflow, packet drop and congestion between routers. NoC architectures use credit-based and shared virtual channel control techniques. At times both these techniques can be used together for controlling access to the shared link. Due to less communication overhead in the shared virtual channel control technique, shared virtual channel control consumes less area and less power compared to the credit-based method (Bjerregaard and Sparsø 2005b).

2.4.1 Credit-Based Technique

In this technique, the flow of flits between routers and PEs is managed by incrementing and decrementing a credit counter. When a port sends a flit, it decrements the credit counter. This specifies that the output port is busy, as it just sends the flit. When the adjacent router receives the flit, it sends back the credit packet to the router so that it increases the credit counter. When the credit counter of the port reaches zero, it cannot send more flits at the particular output port. Figure 2.2 shows the

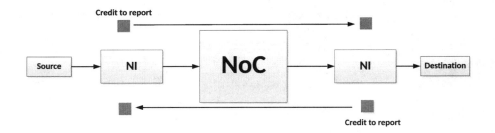

FIGURE 2.2 Credit-based buffer management.

credit-based buffer management scheme. This technique is employed between adjacent routers and network interfaces (NIs) (Hansson, Goossens, and Rădulescu 2007).

The NoC architectures that have credit-based buffer management and flow control techniques are (Guerrier and Greiner 2008; Bolotin et al. 2004; Stefan, Molnos, and Goossens 2012; Bell et al. 2008; Gratz et al. 2007; Peña-Ramos and Parra-Michel 2011; Yang et al. 2014; Evain, Diguet, and Houzet 2004; Feliciian and Furber 2004; Lattard et al. 2008; Lee et al. 2012; Kao and Chao 2012).

2.4.2 Shared Virtual Channel Control Technique

In this technique, there is a share and unshare box counter at the source and destination. The share box is locked whenever a flit is sent from the source queue. After this, the source is not allowed to send more flits until the flit is received at the destination. When the flit is reached at the destination, it toggles the latch of the unshared box, which in return sends the unlock signal to the shared box via the unlock link, as shown in Figure 2.3. In this way, the source can send more flits on the network, provided there is no deadlock (Bjerregaard and Sparso 2005a). The Mango (Bjerregaard and Sparso 2004, 2005a, 2005c) NoC architecture has adopted the shared virtual channel buffer management and flow control technique. In contention-free routing, resources, buffers and router ports (slots) are reserved for a specific time period, so that there is no contention between multiple connections (Rijpkema et al. 2003).

The other flow control techniques are handshaking, acknowledge (ack)/not acknowledge (nack), stall/go signals, go-back-N, separate control lines and header between routers, NIs and PEs. The details are provided in Appendix A.

2.5 NoC EVALUATION PARAMETERS

The NoC architectures are evaluated based on various parameters. These parameters include latency, throughput, bandwidth utilization, area, power dissipation, energy consumption and frequency. Every NoC architecture attempts to maximize the throughput and bandwidth utilization and minimize the latency, area, power and energy dissipation using various techniques. Latency refers to the time taken by a packet or flit to move from the source to the destination. Usually, in NoC, latency is measured in nanoseconds (ns). Throughput is the number of bits that can be sent across the NoC interconnect (link) per second. Throughput is measured in bits per second (bps). Bandwidth is the maximum number of bits that can be sent over a interconnect (link) per second. The unit of bandwidth is bits per second (bps). Area refers

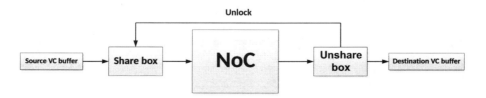

FIGURE 2.3 Shared virtual channel control scheme.

In-Depth Review of Network on Chip

to the size of the NoC after synthesis. Nowadays, there are also some software tools that can measure the area and power consumption of the NoC architecture. The unit of area used by most architectures is mm². The units of energy and power consumption used by most architectures are milliwatt (mW) and petajoules (pJ), respectively. When packets traverse the router and interconnect, they consume power and energy.

These parameters are critical to calculate as more heterogeneous devices (PEs) come onto chips. They increase the power and area consumption of the chip. Appendix A shows the parameters by which different NoC architectures are evaluated.

2.6 NoC CLOCKING MECHANISM

Communication between routers, NIs and PEs are mostly asynchronous, synchronous, globally asynchronous and locally synchronous (GALS), mesochronous, plesichronous and heterochronous. The trends show that the NoC architectures are moving from synchronous communication to asynchronous communication. The reason is that asynchronous communication is faster and activity dependent, exploiting average-case rather than worst-case performance (Amde et al. 2005; Ben-Itzhak et al. 2012; Emerson 1997). Designing asynchronous NoC is modular, but it is more complex compared to synchronous NoC (Emerson 1997). Designing a glitch-free circuit that at the same time manages multiple clocks is more difficult to manage than global clock–based communication. This leads to the idea of synchronous and asynchronous NoC communication by having a GALS architecture. The GALS architecture solves the timing issues in multiple PEs in a NoC because local synchronous PEs need not synchronize with a single global clock. The GALS approach consumes less power compared to the global clock approach, as local PEs have more control over their clocks independently from other synchronous PEs (Agarwal, Iskander, and Shankar 2009). Depending on the requirements of the application mapped on the NoC, the synchronous, asynchronous or GALS architecture will support it and will increase the performance of the NoC. The clock frequency of the NoC architectures is also shown in Appendix A. The frequency is measured in hertz (Hz). The scale of the frequency used by the NoC architectures is megahertz (MHz) and gigahertz (GHz).

2.6.1 SYNCHRONOUS NoC

In synchronous NoC, communication and activities between multiple components are synchronized based on a single centralized clock. The benefit of synchronous communication is that all of the changes at the routers, PEs and links occur at the same time at a regular clock period. The drawback of this mechanism is clock skew, lack of modularity, electromagnetic interference, worst-case performance and clock power consumption issues (Amde et al. 2005). Examples of synchronous NoCs are (Guerrier and Greiner 2008; Rijpkema et al. 2003; Liang, Swaminathan, and Tessier 2000; Bolotin et al. 2004; Leroy et al. 2005; Hilton and Nelson 2006; Paukovits and Kopetz 2008; Stefan, Molnos, and Goossens 2012; Dally and Towles 2001; Henriksson, Wiklund, and Liu 2003; Bertozzi and Benini 2004; Zeferino, Kreutz, and Susin 2004; Mullins, West, and Moore 2006; Castells-Rufas et al. 2009; Bell et al. 2008; Gratz et al. 2007; Soteriou et al. 2006; Stensgaard and Sparsø 2008;

Samman, Hollstein, and Glesner 2009; Feero and Pande 2008; Lan et al. 2011; Logvinenko, Gremzow, and Tutsch 2013; Postman et al. 2012; Yaghini, Eghbal, and Bagherzadeh 2015; Penolazzi and Jantsch 2006; Millberg et al. 2004; Kavaldjiev, Smit, Jansen et al. 2006; Chi and Chen 2004; Kim, Kim, and Sobelman 2005; Bouhraoua and Elrabaa 2006; Sanusi and Wang 2006; Lee et al. 2006; Butts 2007; Wang et al. 2010; Qouneh et al. 2012).

2.6.2 Asynchronous NoC

In asynchronous NoC, activities are performed based on the handshake (control) signals shared between routers (Bjerregaard and Mahadevan 2006; Tatas et al. 2014; Amde et al. 2005; Yaghini et al. 2009). Asynchronous NoCs are also called self-timed NoCs (Bainbridge and Furber 2002). Asynchronous circuit implementation usually takes more area, delay, power consumption and channel bandwidth compared to the synchronous approach because of the explicit sharing of signals (Amde et al. 2005). It is not easy to resynchronize the asynchronous communication. The resynchronization introduces error bits and increases the latency and power consumption (Bjerregaard and Mahadevan 2006). The asynchronous communication is faster compared to synchronous communication at high traffic loads, as it has the least latency (Bjerregaard and Mahadevan 2006; Ben-Itzhak et al. 2012). In asynchronous NoC, there can be no timing relationship between synchronous clocks (PEs). Asynchronous provides the maximum flexibility in terms of timing (Teehan, Greenstreet, and Lemieux 2007). When interconnects are idle, apart from the leakage, no power is consumed in asynchronous NoC. This solves the problem of increasing power consumption, as the chip size is increased (Bjerregaard and Mahadevan 2006). The examples of asynchronous NoCs are (Moraes et al. 2004; Rijpkema et al. 2003; Chang, Shen, and Chen 2006; Bainbridge and Furber 2002; Pande et al. 2003; Zeferino and Susin 2003; Samuelsson and Kumar 2004; Bartic et al. 2005; Kumar et al. 2008; El-Moursy, Korzec, and Ismail 2009; Krasteva, De la Torre, and Riesgo 2010; Wu, Tang, and Hsu 2011; Peña-Ramos and Parra-Michel 2011; Mishra, Nidhi, and Kishore 2012; Shu et al. 2012; Al Faruque, Ebi, and Henkel 2010; Choudhary and Qureshi 2012; Evain, Diguet, and Houzet 2004; Coppola et al. 2004; Feliciian and Furber 2004; Bjerregaard and Sparso 2004, 2005a, 2005c; Wolkotte et al. 2005; Beigné et al. 2005; Mondinelli, Borgatti, and KOVACS VAJNA 2004; Ching, Schaumont, and Verbauwhede 2004; Amde et al. 2005; Lattard et al. 2008; Lee et al. 2012).

2.6.3 Globally Asynchronous Locally Synchronous NoC

In the GALS approach, the NoC is divided among multiple subsystems. Each subsystem has its own clock, and they communicate with other systems through asynchronous NoC communication. GALS has less power consumption compared to using a global clock. It avoids the problem of clock synchronization due to clock skew, as more and more devices are coming onto the chip (Amde et al. 2005; Teehan, Greenstreet, and Lemieux 2007; Sigüenza-Tortosa, Ahonen, and Nurmi 2004). Figure 2.4 shows the GALS architecture where independent synchronous islands of systems are

In-Depth Review of Network on Chip

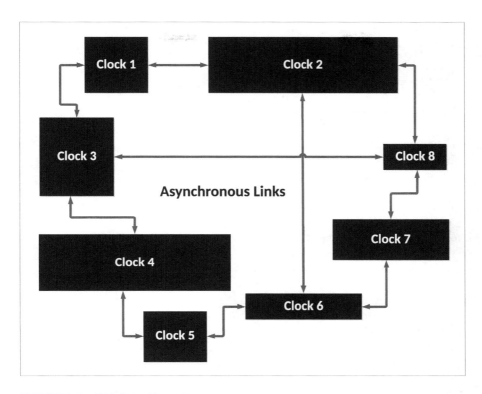

FIGURE 2.4 GALS clocking scheme.

connected with asynchronous NoCs through irregular topologies. Examples of NoC architectures that follow the GALS approach are (Lines 2004; Castells-Rufas, Joven, and Carrabina 2006; Wang, Ahonen, and Nurmi 2007; Janarthanan and Tomko 2008; Hansson, Subburaman, and Goossens 2009; Siguenza-Tortosa and Nurmi 2002; Kumar et al. 2002; Janarthanan, Swaminathan, and Tomko 2007; Göhringer et al. 2010; Yang et al. 2014; Arteris 2005; Beigné et al. 2009; Liang et al. 2004; Lee et al. 2004; Schoeberl 2007).

In the mesochronous scheme, all the subsystems have different clocks with the same frequency but the phase is different (Wiklund and Liu 2003). This avoids the problem of clock skew. The examples of mesochronous NoC architectures are (Rijpkema et al. 2003; Lee, Lee, and Yoo 2006; Vangal et al. 2008; Wiklund and Liu 2003; Panades, Greiner, and Sheibanyrad 2006; Lee et al. 2005). In plesichronous NoC, the source and destination PE operate at the nominal frequency, which may be slightly different, and this leads to a drifting phase (Teehan, Greenstreet, and Lemieux 2007). The star connected OCN (Lee et al. 2003) is a plesichronous NoC. In a heterochronous NoC, the source and destination have totally different clock frequencies (Teehan, Greenstreet, and Lemieux 2007). Some of the NoC architectures have not been mentioned, or it is not clear which clocking scheme they are using. Further details can be found in Appendix A.

2.7 NoC TOPOLOGIES

The topology of the NoC defines how the PEs, routers and links are connected with each other in the network. The selection of a particular topology for any specific NoC depends on the communication usage of the PEs. Other factors in choosing a particular topology for any NoC design include its impact on performance and cost parameters. The performance parameters include latency, fault tolerance, bandwidth and throughput, while cost parameters include area and power consumption (Ogras, Hu, and Marculescu 2005; Stefan, Molnos, and Goossens 2012). Many different topologies have been proposed (Wang et al. 2011), and there are broadly two categories: regular and irregular topologies. In regular topologies, routers and PEs are connected according to some patterns, for example, mesh, torus, fat tree, butterfly fat tree, octagon, star, hierarchical star, crossbar, ring, spidergon, chain, subnets and generalized de Bruijn topology. However, there is no standard pattern in an irregular topology, which may include hybridized topologies. Although regular NoC topologies are simple to implement, at times they have the drawback of being nonoptimal in terms of network utilization. The low network utilization leads to delayed communication and increased power consumption. Irregular topologies are used in the literature to overcome these drawbacks. The irregular or custom NoC topologies are application-specific and mostly have heterogeneous PEs and routers compared to regular NoC topologies (Tatas et al. 2014). Regular topologies make the routing decision simple, and it is easily replicated on multiple nodes, while irregular topologies make the routing algorithm more complex (Guerrier and Greiner 2008).

In the mesh topology, the routers are connected to each other through a point-to-point connection in a mesh structure, as shown in Figure 1.1. The routers at the first and last column or row of the NoC are not connected with any other neighboring routers. Most of the NoC architectures have a mesh topology because its communication structure is less complicated and it is also similar to a 2D silicon surface (Seifi and Eshghi 2012; Teehan, Greenstreet, and Lemieux 2007). One of the drawbacks of the mesh topology is congestion at the center of the NoC due to the routing algorithm (Tatas et al. 2014). In a mesh topology, the routing algorithm should have an equal and balanced traffic distribution mechanism. The torus topology is similar to the mesh except that the first and last column or row elements are also connected with each other, as this increases and eases the routing decisions. The torus topology solves the excessive round end trip delays, as they are connected together. The drawback of the torus topology is more area and links requirement than the mesh NoC (Wang et al. 2011). Figure 1.3 and Figure 1.4 show the torus and irregular NoC topologies. In tree structure topologies, the routers are connected in a hierarchical design such that the parent's routers have child routers connected with them. The PEs are connected to leaf routers, as shown in Figure 1.3. The tree topology is hierarchical in structure but becomes complex as more PEs are joined at the leaf routers (Guerrier and Greiner 2008). In tree-based topology, the traffic can be quickly disseminated to the desired destination due to the hierarchical structure (Bjerregaard and Mahadevan 2006).

In the star topology, the routers are connected with a centralized arbiter. The arbiter manages communication between multiple routers. The arbiter may be a specialized device, or it can be a router as well (Lee et al. 2003). The capacity of the centralized arbiter/router should be significant to manage the traffic received from the multiple PEs connected to it. At times this also leads to congestion at the centralized switch (Tatas et al. 2014). In the octagon topology, however, eight routers are connected in a ring fashion with the bidirectional wires (Karim, Nguyen, and Dey 2002). The octagon topology is simple to implement. Fast and efficient routing algorithms can be implemented using this topology. Excessive wiring is required to extend the octagon network to more than eight PEs (Tatas et al. 2014). In the crossbar topology, the routers are connected with several wires in a crossbar fashion. This topology is simple to implement, but the links between PEs become complex and broad as more PEs are connected (Lines 2004). Ring topology refers to the nodes connected in a ring (Samuelsson and Kumar 2004), and all routers are connected with each other through interconnects in a ring shape, as shown in Figure 1.4. Although the ring topology is simple to implement, it has limited scalability and performance problems as the number of nodes is increased (Tatas et al. 2014).

In the spidergon topology, several nodes (more than two) are connected together in a bidirectional ring in both clockwise and counterclockwise directions. In this way, the nodes in the spidergon are connected with a cross-connection. The spidergon topology is similar to a spider web. This topology is regular, scalable, point-to-point and has a suitable network diameter. The network becomes complex as the number of PEs is increased (Coppola et al. 2004). In the subnet topology, small groups of PEs are connected with the router, which are called the subnets. These subnets in turn are connected to each other. The PEs that communicate often are put in the same subnet. This makes communication between PEs more efficient. The communication between PEs in different subnets takes more time, which may lead to delayed communication (Hilton and Nelson 2006).

In the de Bruijin topology, the nodes are connected through vertices, and it can be an irregular topology (Soteriou et al. 2006). It is, however, an efficient topology for parallel processing and is also suitable for very large-scale integration (VLSI) implementation. The two-dimensional de Bruijin NoC performs better compared to the two-dimensional mesh topology in terms of latency and energy dissipation. At times, in two-dimensional de Bruijin the NoC takes more links to connect the neighbor node compared to the two-dimensional meshes, and this increases the network area (Sabbaghi-Nadooshan, Modarressi, and Sarbazi-Azad 2010). The details of NoC architectures that have a specific topology are given in Appendix A.

2.8 OPEN SOURCE

The open-source software/architecture has a license allowing the end user to have access to the source code, redistribute the software/architecture and modify the original code with the same license. At times, the license may have special provisions. For example, the derived software/architecture should use its own name and version apart from mentioning the original software/architecture name. Open-source architectures and codes are commonly provided for academic and research purposes. They are mostly not allowed to be used for commercial purposes.

Æthereal (Rijpkema et al. 2003), Aelite (Hansson, Subburaman, and Goossens 2009) and dAElite (Stefan, Molnos, and Goossens 2012) architectures are not open source. However, Nostrum (Penolazzi and Jantsch 2006; Millberg et al. 2004) is provided on request. Reconfigurable Network on Chip (Ching, Schaumont, and Verbauwhede 2004) and NocMaker (Castells-Rufas et al. 2009) are open-source architectures. The specific details are given in Appendix A.

2.9 NoC CONNECTION TYPES

Connection type refers to the way a source PE is connected with the destination PE. The connections can be simple, narrowcast, multicast or broadcast. The multicast communication produces a significant amount of redundant traffic that increases latency and congestion in the NoC compared to unicast communication. By combining multiple unicast connections, multicast communication can be constructed to overcome the drawbacks of multicast communication. Various multiprocessor System on Chip (MPSoC) applications use multicast communication, which includes replication, barrier synchronization, and cache coherency in a distributed shared memory architecture (Palesi and Daneshtalab 2014). Figure 2.5 shows the three connection types.

2.9.1 SIMPLE CONNECTION

1–1, or simple connection, is between one source PE and one destination PE (Radulescu and Goossens 2004). A few architectures that provide simple communication between PEs are (Guerrier and Greiner 2008; Moraes et al. 2004; Rijpkema et al. 2003; Liang, Swaminathan, and Tessier 2000; Karim, Nguyen, and Dey 2002; Lines 2004; Hansson, Subburaman, and Goossens 2009; Stefan, Molnos, and Goossens 2012; Bainbridge and Furber 2002; Pande et al. 2003; Henriksson, Wiklund, and Liu 2003; Bertozzi and Benini 2004; Samuelsson and Kumar 2004; Castells-Rufas et al. 2009; Gratz et al. 2007; Vangal et al. 2008; Kumar et al. 2008; El-Moursy, Korzec, and Ismail 2009; Göhringer et al. 2010; Krasteva, De la Torre, and Riesgo 2010; Wu, Tang, and Hsu 2011; Peña-Ramos and Parra-Michel 2011; Grot et al. 2011; Mishra, Nidhi, and Kishore 2012; Shu et al. 2012; Logvinenko, Gremzow, and Tutsch 2013; Arteris 2005; Wiklund and Liu 2003; Penolazzi and Jantsch 2006; Millberg et al. 2004; Bjerregaard and Sparso 2004, 2005a, 2005c; Wolkotte et al. 2005; Kavaldjiev, Smit, Jansen et al. 2006; Panades, Greiner, and Sheibanyrad 2006; Liang et al. 2004; Hu and Marculescu 2004; Schoeberl 2007; Lee et al. 2012; Bahirat and Pasricha 2014).

2.9.2 NARROWCAST CONNECTION

A connection becomes narrowcast if one source PE is connected with one or multiple destinations PEs. In this connection, the instruction initiated by the source is only executed by one destination PE. The destination can send a return message to the source PE to acknowledge the instruction or data sent by the source PE. Narrowcast connections are bidirectional between the source and destination (Radulescu and Goossens 2004).

In-Depth Review of Network on Chip

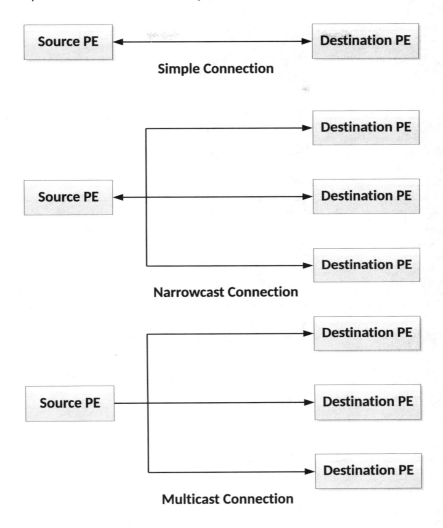

FIGURE 2.5 Simple, narrowcast and multicast connections.

2.9.3 Multicast Connection

These connections are between a source PE and one or more destinations PEs. In these connections, instructions or data sent by the source are duplicated, and the copies are sent to every destination PE. As these connections are unidirectional, no return messages are allowed from the destination PE to the source PE. Return messages are also not allowed due to memory limitations. Multicast connections are unidirectional (Radulescu and Goossens 2004). The examples of NoC architectures providing multicast connections are (Rijpkema et al. 2003; Salminen et al. 2006; Wang, Ahonen, and Nurmi 2007; Hansson, Subburaman, and Goossens 2009; Stefan, Molnos, and Goossens 2012; Bainbridge and Furber 2002; Pande et al. 2003;

Kumar et al. 2008; Samman, Hollstein, and Glesner 2009; Krasteva, De la Torre, and Riesgo 2010; Wu, Tang, and Hsu 2011; Grot et al. 2011; Ganguly et al. 2010; Penolazzi and Jantsch 2006; Millberg et al. 2004; Bjerregaard and Sparso 2004, 2005a, 2005c; Schuck, Lamparth, and Becker 2007; Liang et al. 2004; Kao and Chao 2012).

2.9.4 BROADCAST COMMUNICATION AND MULTIPATH ROUTING

As the names imply, with these types of communication, a source PE can send packets to every destination PE attached with it directly or indirectly (Guerrier and Greiner 2008). SPIN (Guerrier and Greiner 2008), Nexus (Lines 2004) and TTNoC (Schoeberl 2007) are a few architectures that support broadcast communication. Multipath routing uses multiple paths available to route packets from source to destination. CLICHÉ (Kumar et al. 2002), on the other hand, provides multipath routing. The specific details can be found in Appendix A.

2.10 NoC SIZE

NoC architectures have used a number of different NoC sizes for implementation purposes. It is because NoC is modular, scalar and the implementation on one NoC size can easily be applied to different NoC sizes. These sizes include 2×2, 3×3, 4×4, 5×5, 6×6, 7×7, 8×8, 5×8, 2×3 and 4×3, having two-dimensional mesh and torus topologies. Some of the NoC topologies are regular, while others are irregular. The details are provided in Appendix A about NoC architectures supporting particular NoC sizes.

2.11 NoC IMPLEMENTATION PLATFORMS

The NoC architectures are implemented using the simulator, field programmable gate array (FPGA) and application-specific integrated circuits (ASIC). Using a cycle accurate simulator, some NoC architectures have been synthesized. The simulators used are OMNET++ (Varga and Hornig 2008), NS-2, OPNET, Noxim, Nirgam (Jain et al. 2007), Simics, Garnet (Agarwal et al. 2009), Orion (Wang et al. 2002), Spice, Recsim (Logvinenko and Tutsch 2012) and CACTI (Muralimanohar, Balasubramonian, and Jouppi 2007). The details of NoC architectures using a specific simulator can be found in Appendix A.

2.12 NoC BUFFERING MECHANISMS

A buffer is an integral part of any network router (Bjerregaard and Mahadevan 2006), but it consumes more power in the NoC. Therefore, an efficient buffer design is essential for optimized NoC performance (Tatas et al. 2014). The design includes the location and size of the buffer (Bjerregaard and Mahadevan 2006). To minimize the access latency and implementation cost, it is always preferred to use registers rather than huge memories in the form of static random access memory (SRAM) or

In-Depth Review of Network on Chip

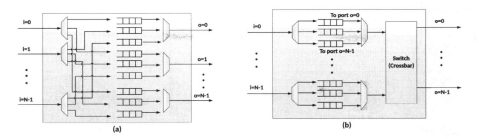

FIGURE 2.6 (a) Output queue and (b) VOQ router architecture.

dynamic random access memory (DRAM) (Hu and Marculescu 2003). The location of the buffers can be at the input or output port of the router. These buffers are called the input queue and output queue. The performance of the output buffer is the best compared to other strategies, but at the cost of more loads on interconnects. When to read the flits from the input or output queue depends on the Scheduler, which in turn checks the output port to see whether it is free. Usually, routers used to have one queue per port, but due to the Head of Line (HOL) problem, now they have more than one queue per port for every connection. The virtual output queue (VOQ) is one type of input buffering that has the combined benefits of input and output queuing (Rijpkema et al. 2003). The input queues may have less memory cost, as they have fewer queues compared to VOQ and output buffers (Radulescu and Goossens 2004). Input queues also solve the problem of HOL, as the output queues may be blocked due to congestion or faulty interconnects (Bjerregaard and Mahadevan 2006). Figure 2.6 shows the output and VOQ architectures of routers. Most of the NoC architectures prefer input and a combined input and output buffers because of the better link utilization and performance. The area of the NoC increases directly as the size of the input buffers is increased. In (Ogras, Hu, and Marculescu 2005), it is mentioned that the 4 × 4 NoC area increases by 30% as the input buffer size is increased from two to three words. But at certain times, depending on the network load, the network latency is reduced in a considerable amount by increasing the buffer size. However, due to the heterogeneous traffic generated by PEs in the NoC, it is always desirable to allocate more buffers for high-traffic channels, whereas a small number of buffers can be allocated to low-traffic channels. Specific details on NoC architectures can be found in Appendix A.

2.13 NoC PE–ROUTER INTERFACE

The interface between the PE and router through the NI either is through a bidirectional bus or follows a particular standard interface. The standard interfaces are AMBA AHB, AMBA APB, AMBA AXI, VCI, PVCI, BVCI, AVCI and OCP that follow a particular procedure and signals for reading and writing data from the source PE and destination PE. Appendix A shows the NoC architectures that have adopted these interfaces.

2.13.1 ADVANCED MICROCONTROLLER BUS ARCHITECTURE

Advanced Microcontroller Bus Architecture (AMBA) is an on-chip interconnect specification. The AMBA family (1, 2, 3, 4, 5) standard helps to manage the connections and data movement between PEs in the NoC.

AMBA defines three types of buses for high-performance communication. These are the advanced system bus (ASB), advanced high-performance bus (AHB) and an advanced peripheral bus (APB). ASB supports multiple bus masters, high burst and pipelined transfer operations (Mitić and Stojčev 2006b). AHB is for those system modules that work at high frequency and have high bandwidth communication requirements. AHB supports multiple bus master operations, split transactions, peripheral and burst transfer and nontristate implementation. APB is for those devices that operate at low power, and it connects general-purpose low-speed devices (Mitić and Stojčev 2006b; Specification 1999). All devices connected with the peripheral bus master (bridge) are called slaves. APB is a static bus, and it supports simple addressing. Figure 2.7 shows that AMBA buses are hierarchically organized in two bus segments. System and peripheral buses are connected together through a bridge that sends data between these two segments. AMBA buses do not define any arbitration rules; instead, the arbiter can decide the application needs (Mitić and Stojčev 2006b).

The AMBA advanced extensible interface (AXI) protocol is for high-performance and high frequency systems. It is an advanced version of AHB, ASB and

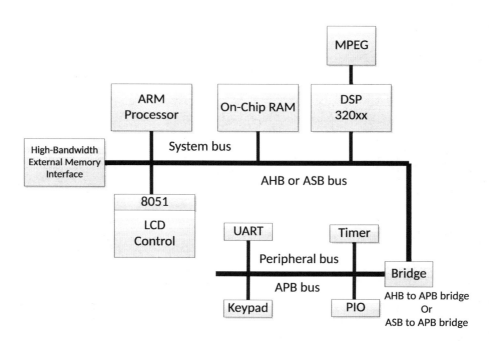

FIGURE 2.7 AMBA buses in a system.

In-Depth Review of Network on Chip

APB, and belongs to the AMBA 2 family. AMBA has also released the latest version of AXI in the AMBA 3 protocol family. AXI is based on point-to-point connections (Mitić and Stojčev 2006b). The AMBA advance trace bus (ATB) is part of the AMBA 4 family released in 2012, and it spelled out the rules of how traces are shared in a trace system between its modules (Benini and Bertozzi 2005). Finally, the Coherent Hub Interface (CHI) is the new SoC protocol in the AMBA 5 family. CHI provides the better performance required by servers and networking applications (ARM). Every family of AMBA is better than the previous one and provides more enhanced features.

2.13.2 Open Core Protocol

Open Core Protocol (OCP) is an interface standard that connects PEs to an on-chip bus. It is an openly licensed protocol. It defines the rules of communication between two modules. The two modules can be a PE and an NI (Tatas et al. 2014). OCP supports a master–slave interface with unidirectional signals. It is fully synchronous and driven by the rising edge of the clock. It supports data transfer on every clock cycle. In OCP, signals are point-to-point except the clock and reset signals. OCP allows the master and slave to control the transfer rate (Mitić and Stojčev 2006a). The drawback of OCP is that Altera and Xilink, the two largest FPGA vendors, do not support this protocol. Figure 2.8 shows the SoC design based on the OCP protocol. The NoC architectures with OCP interfaces are (Moraes et al. 2004; Chang, Shen, and Chen 2006; Salminen et al. 2006; Wang, Ahonen, and Nurmi 2007; Wingard 2001; Bainbridge and Furber 2002; Zeferino and Susin 2003; Bertozzi and Benini 2004; Hoffman et al. 2007; Arteris 2005; Bjerregaard and Sparso 2004, 2005a, 2005c; Panades, Greiner, and Sheibanyrad 2006).

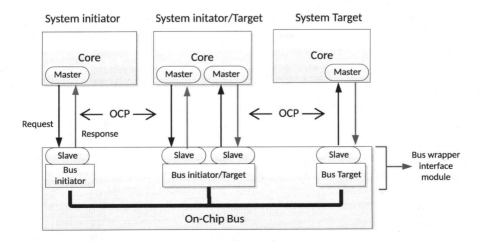

FIGURE 2.8 SoC design using OCP protocol.

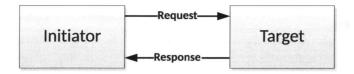

FIGURE 2.9 Point-to-point connection in VCI.

2.13.3 VIRTUAL SOCKET INTERFACE ALLIANCE (VSIA)

PROTEO (Siguenza-Tortosa and Nurmi 2002) follows the standard VSIA-compliant interface, which includes three different virtual component interfaces (VCIs). They are peripheral VCI (PVCI), basic VCI (BVCI) and advanced VCI (AVCI). VCI is an interface, and it specifies the request–response protocol, a set of instructions for transferring request–response messages and the content of these messages. PVCI is a simple interface and is easily implementable compared to BVCI. BVCI is suitable for most applications. It is a very powerful protocol, but not very complex. AVCI is an advanced version of BVCI with more advanced features of the threads. AVCI supports high-performance applications. Figure 2.9 shows a point-to-point connection between the initiator and target using VCI. The initiator issues a request, while the target replies with a response signal. VCI can be connected to a bus using an interface wrapper, as shown in Figure 2.10. The initiator is connected via a VCI interface to the bus initiator. The bus initiator wrapper is connected with the bus. Similarly, on the target side, the bus wrapper is used. At the initiator bus wrapper, the VCI target and bus master are used, while at the target side, a bus slave and VCI initiator are used (Mitić and Stojčev 2006a). Examples of NoC architectures that have VCI interfaces are (Guerrier and Greiner 2008; Wang, Ahonen, and Nurmi 2007; Zeferino and Susin 2003; Evain, Diguet, and Houzet 2004; Panades, Greiner, and Sheibanyrad 2006; Lee, Lee, and Lee 2004).

According to the literature, most of NoC architectures have not mentioned the type of interface. Only twelve architectures use OCP interfaces, while seven use the VSIA interface. The AMBA family interfaces are adopted by three NoC architectures. Other on-chip buses are Avalon, CoreConnect, STBus, Wishbone, CoreFrame, Marble, PI Bus and the SiliconBackplane network (Mitić and Stojčev 2006a).

2.14 NoC FREQUENCY AND TECHNOLOGY

The NoC frequency refers to the clock frequency of the NoC. There are different clocking mechanisms based on which NoC is synchronized. They are asynchronous, synchronous, GALS, mesochronous, plesichronous and heterochronous (Teehan, Greenstreet, and Lemieux 2007). The frequency is measured in Hz. The scale of the frequency used by the NoC architectures is measured in MHz and GHz. The frequency of the NoC architecture specifies how quickly it can process the instructions and data. Appendix A specifies the frequency parameters of NoC architectures. Most of the NoC architectures have specified the frequency at which they operate, but only thirty-three architectures do not specify the clock frequency.

In-Depth Review of Network on Chip

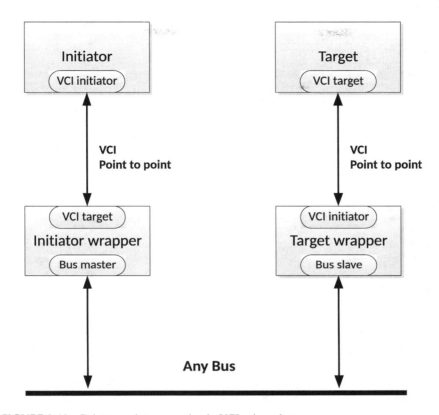

FIGURE 2.10 Point-to-point connection in VCI using a bus.

Technology refers to the feature size. Feature size is the minimum size of a transistor or a wire. NoC architectures are implemented in micrometers (μm) and nanometer (nm) technologies. Forty-three percent of the NoC architectures have not mentioned the technology on which their NoC architectures are implemented.

These features might help the research community to shortlist the NoC architecture based on the frequency and feature size, depending on their specific application requirements and do further research on it.

2.15 NoC AREA AND POWER CONSUMPTION

Area refers to the size of the NoC on the system on a chip, on an FPGA or on ASIC. Some of the NoC architectures have mentioned the area of the complete NoC, whereas a few of them have specified the area of the router or switch. The area is usually measured as the total number of lookup tables or slices used in FPGA or the number of gates/transistors/flip-flops/logic cells used in the NoC architecture. Some of the NoC architectures have specified the area of the NoC in mm^2 and μm^2. The details regarding specific NoC architectures' areas are presented in Appendix A. In embedded systems, the area and power consumption of the devices are vital.

It is an open research area for developing techniques to efficiently utilize these two parameters.

The NoC clocking mechanism, buffer management, routing algorithms and switching activities consume power and energy on a per-router and interconnect basis. The details regarding these parameters related to specific NoCs are provided in Appendix A. The units used for power dissipation are µW (microwatt), mW (milliwatt), W and MW (megawatt), while the units used for energy consumption are pj (petajoule) and pj/packet (petajoule per packet).

2.16 NoC ROUTER PORTS AND BUS WIDTH

The number of ports in the switch improves the performance of the NoC, but it might also increase the complexity of the NoC. The number of ports in the switch also depends on the topology being used in the NoC. The area consumption of the router increases as the number of ports is increased (Tatas et al. 2014). The number of ports and topology should be optimally selected to improve the overall performance of the NoC. For a mesh topology, five ports are connecting four immediate neighbors and one to the local core (PE). For tree and other topologies, it can be a different number of ports based on the NoC configuration. In 3D architectures, two more ports are connecting the upper and lower layers of the NoC. QORE (DiTomaso, Kodi, and Louri 2014) has the highest number of router/switch ports (i.e., twenty), while the minimum number of ports is three. The details of other NoC architectures can be found in Appendix A. The number of router ports has a significant effect on the topology of the NoC. This, in turns, affects the bandwidth, throughput, latency and power consumption of the NoC. Information regarding the number of ports is essential for the NoC. Sixty-three percent of NoC architectures have provided information regarding the router ports.

Bus or channel width is the number of bits that can be sent over the channel. It controls the number of bits that can be sent over the channel from multiple sources and destinations. This parameter affects the congestion, deadlock, bandwidth and throughput utilization of the NoC. Usually, the bus width is equal to the flit size. Bus width also affects the complexity, latency, power and area consumption of the NoC (Tatas et al. 2014; Marculescu et al. 2008; Brugge and Khalid 2014). To have an optimal channel width at times is not possible, as it is dependent on the application and other factors involved in the NoC (Marculescu et al. 2008). Different NoC architectures have a different channel and data bus width. Only 45% of NoC architectures have provided details regarding bus or channel width. Details about the channel width of different NoC architectures are given in Appendix A.

2.17 NoC YEAR OF PROPOSAL, FLIT SIZE AND LATENCY

The NoC architectures have been proposed from 2000 to 2019. Some of the architectures are big and well defined, while others are used for particular applications. In 2000, the first NoC architecture SPIN (Guerrier and Greiner 2008) was proposed. This NoC architecture supported synchronous packet switching. It had wormhole switching and credit-based buffer management technique for flow control and buffer

In-Depth Review of Network on Chip

management. It possessed a fat tree topology and enabled 1–1 and broadcast communication. The SPIN (Guerrier and Greiner 2008) NoC architecture laid the foundation for further research in the area of on-chip network communication. Later, in 2001, Dally et al. (Dally and Towles 2001) proposed the on-chip network and router architecture. In 2002, Luci Benini named this on-chip network "Network on Chip" (Benini and De Micheli 2002). That year was the cornerstone for further research in the NoC area. Research flourished on the different aspects of NoC, which include different topologies, routing algorithms, buffer management, quality of service (QoS) parameters, clocking mechanisms, fault tolerance, switching techniques, congestion avoidance and management techniques, multiple connection types and other areas. Appendix A shows the year-based information for the various NoC architectures.

Flit stands for "flow control digits." It is the smallest unit of a packet. Packets are divided into a header, body and tail flit for efficient routing, bandwidth and throughput utilization. The size of the flit affects the latency, channel (bus) width and power consumption of the NoC. Appendix A shows the different NoC architectures that have different flit sizes. Flit size is commonly measured in bits. Some of the NoC architectures have the packet size in flits and bits. The message generated by the source PE contains packets, which are divided into flits. By keeping the flit size the same and shrinking the size of the feature as technology scales down, the overall area of the router is decreased but power consumption is increased. Therefore, technology scaling does not allow the width of the flit to be increased. The cost (area and power consumption) of the router increases along with the flit size. The NoC saturation point is reached early as the flit size is continuously increased. The general rule of thumb is to peg the size of the flit to the size of the shortest packet. To increase the throughput of the NoC, one way is to increase the flit size, but this is not a cost-effective solution, as it may lead to an issue of fragmentation (Lee et al. 2013). The flit size in HERMES (Moraes et al. 2004) and AdNoC (Al Faruque, Ebi, and Henkel 2010) is 8 bits. The flit size is 16 bits for Dally et al. (Dally and Towles 2001), CHAIN (Bainbridge and Furber 2002), QNoC (Bolotin et al. 2004), On-Chip Multimedia Applications (Lee et al. 2006) and BMNoC (Lee et al. 2012) NoC architectures. The following NoC architectures (Guerrier and Greiner 2008; Liang, Swaminathan, and Tessier 2000; Bertozzi and Benini 2004; Hoffman et al. 2007; Arteris 2005; Beigné et al. 2009; Hu and Marculescu 2004) have 32-bit flits. Similarly, the flit size is 34 bits for A Reconfigurable Baseband Platform Based on Asynchronous NoC (Lattard et al. 2008) and ReNoC (Stensgaard and Sparsø 2008). In Intel TeraFLOPS (Vangal et al. 2008), the flit size is 38 bits. The flit size is 42 bits for BIDI-MIN (Pande et al. 2003). For Polaris (Soteriou et al. 2006) and BLOCON (Kao and Chao 2014), the flit size is 64 bits. In MicroNetwork (Wingard 2001) the flit size is 128 bits, while HELIX (Bahirat and Pasricha 2014) has a flit size of 256 bits. In CLICHÉ (Kumar et al. 2002), the packet size is 256 bits, whereas Æthereal (Rijpkema et al. 2003) has a flit size of 8 flits.

Latency is the time required to send a flit from the source PE to destination PE. The latency of the flit and packet depends on the time taken by the router and interconnects to deliver it to a neighboring router or interconnect. Latency also depends on the flit size, channel or bus width, topology and communication standards of NoC (Marculescu et al. 2008). In Appendix A, latency is measured in cycles, nanoseconds (ns), microseconds (µs) and picoseconds (ps) in different NoC architectures. Most of these NoC architectures have not provided this information. Forty-two percent of NoC architectures have specified latency information. The details are mentioned in Appendix A.

2.18 QUALITY OF SERVICE

Guaranteed throughput/guaranteed service (GT/GS) and best-effort services (BE) are two broad categories of services provided in the NoC (Nurmi 2005). The literature shows that most of the NoC architectures offer BE or packet-based quality of service (QoS). Only a few NoC architectures offer GT/GS or both QoS.

2.18.1 GUARANTEED THROUGHPUT SERVICES

In GT connections, resources are reserved for a particular time period between a specific source and destination pair. The resources include channel bandwidth, routers and PE time slots. GT connections may underutilize network resources at certain times. This makes the GT connections an expensive option. At certain times, routers and PEs send a burst of data on these GT connections and then remain silent for a certain period of time. This leads to the underutilization of network resources. That is why BE services complement the GT services by utilizing the unused bandwidth. Video processing is an example of a GT connection. This implies that GT connections are usually preferred for real time-critical traffic applications (Goossens, Dielissen, and Radulescu 2005; Rijpkema et al. 2003). Examples of the NoC architectures that provide GT-based communication services are (Liang, Swaminathan, and Tessier 2000; Karim, Nguyen, and Dey 2002; Lines 2004; Bolotin et al. 2004; Leroy et al. 2005; Hilton and Nelson 2006; Chang, Shen, and Chen 2006; Salminen et al. 2006; Castells-Rufas, Joven, and Carrabina 2006; Wang, Ahonen, and Nurmi 2007; Paukovits and Kopetz 2008; Janarthanan and Tomko 2008; Hansson, Subburaman, and Goossens 2009; Stefan, Molnos, and Goossens 2012).

2.18.2 BEST-EFFORT SERVICES

Bandwidth is not reserved in BE connections. BE uses the bandwidth unused by the GT connections. Thus, the BE connection does not provide any guarantee of bandwidth. BE connections efficiently utilize the bandwidth and resources, as they are designed for average-case scenarios compared to the GT worst-case mechanism. Examples of BE connections are cache updates. This shows that BE connections are preferred for noncritical traffic (Goossens, Dielissen, and Radulescu 2005; Rijpkema et al. 2003). One of the major drawbacks of the BE services is its unpredictability (Tatas et al. 2014). Examples of NoC architectures that provide BE-based communication services are (Guerrier and Greiner 2008; Moraes et al. 2004; Wingard 2001; Bainbridge and Furber 2002; Siguenza-Tortosa and Nurmi 2002; Kumar et al. 2002; Taylor et al. 2002; Pande et al. 2003; Henriksson, Wiklund, and Liu 2003; Zeferino and Susin 2003; Bertozzi and Benini 2004; Samuelsson and Kumar 2004; Zeferino, Kreutz, and Susin 2004; Chan and Parameswaran 2004; Bartic et al. 2005; Kim et al. 2005; Neeb and Wehn 2008; Kariniemi and Nurmi 2006; Mullins, West, and Moore 2006; Zid et al. 2006; Lee, Lee, and Yoo 2006; Castells-Rufas et al. 2009; Bell et al. 2008; Gratz et al. 2007; Vangal et al. 2008; Hoffman et al. 2007; Janarthanan, Swaminathan, and Tomko 2007; Hosseinabady et al. 2007; Soteriou et al. 2006; Kumar et al. 2008; Stensgaard and Sparsø 2008; El-Moursy, Korzec, and Ismail 2009; Samman, Hollstein, and Glesner 2009; Feero and Pande 2008; Lan et al. 2011; Göhringer et al. 2010; Krasteva, De la Torre, and Riesgo 2010; Wu, Tang, and Hsu 2011; Peña-Ramos and Parra-Michel 2011; Grot et al. 2011; Mishra, Nidhi, and Kishore 2012; Shu et al. 2012; Yang et al. 2014; Al Faruque, Ebi, and Henkel 2010; Choudhary and Qureshi 2012; Logvinenko, Gremzow, and Tutsch 2013; Postman et al. 2012; Arteris 2005; Ganguly et al. 2010; Yaghini, Eghbal, and Bagherzadeh 2015).

2.18.3 COMBINED BE-GT SERVICES

The router handles both the BE and GT services connections at the NoC. There is an arbitration unit at the router, which separates the two. The router reads the packet format, which has a bit pattern or field that specifies whether this packet should be sent on the BE or GT connection. These packet bits are set by the PE that generates the traffic (Rijpkema et al. 2003; Dielissen et al. 2003). Another approach to provide BE and GT services in the NoC is by using virtual channels. The high-priority virtual channels are used for GT traffic, while the low-priority virtual channels are assigned to BE traffic (Kavaldjiev, Smit, Wolkotte et al. 2006; Dobkin, Ginosar, and Cidon 2007). Some NoC architectures provide both BE- and GT-based communication; those are (Rijpkema et al. 2003; Wiklund and Liu 2003; Penolazzi and Jantsch 2006; Millberg et al. 2004; Evain, Diguet, and Houzet 2004; Coppola et al. 2004; Feliciian and Furber 2004; Bjerregaard and Sparso 2005a, 2004; Wolkotte et al. 2005; Beigné et al. 2005; Kavaldjiev, Smit, Jansen et al. 2006; Panades, Greiner, and Sheibanyrad 2006; Schuck, Lamparth, and Becker 2007; Beigné et al. 2009; Wang and Bagherzadeh 2014; Heisswolf et al. 2015).

Figure 2.11 shows the mechanism of BE and GT connections. In Figure 2.11(a) the bandwidth of the NoC is not adequately utilized by the GT connections. The white space shows the unused bandwidth of the NoC. BE connections are utilizing this unused bandwidth, as shown in Figure 2.11(c). Figure 2.11(b) shows the average bandwidth requirement of BE connections, where r_{GT} is the resource allocation for the worst-case (GT), r_{BE} is the resource allocation for BE services and r_{AVG} is the resource allocation for the average requirement over time.

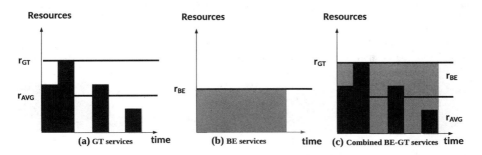

FIGURE 2.11 BE and GT services mechanism.

SUMMARY

Due to the advanced nanoscale manufacturing processes, the communication between the PEs faces various faults on the NoC. The communication requirement in the NoC has increased due to more complex connections. The complexity of NoC interconnects is due to the physical device sizes being scaled down. Various fault-tolerant routing algorithms have been proposed. The fault-tolerant algorithms are deterministic, stochastic, fully adaptive and partial adaptive routing algorithms. However, all of these algorithms do not entirely address the issues of fault tolerance. Biological brain-inspired algorithms have been adapted to make NoCs fault-tolerant and robust. The bio-inspired "synaptogenesis" and "sprouting" algorithms are a self-adopting and self-healing concept from the biological brain. These algorithms were implemented with BE, GT and combined BE-GT services. Two survey studies were conducted to gain in-depth knowledge of the NoC architecture. The NoC architectures were evaluated using different parameters, among which fault tolerance was one of the essential parameters.

REFERENCES

AMBA® 3 AXI™ Protocol Checker.

AMBA® AXI Protocol Specification, ARM.

AMBA™ 4 ATB Protocol Specification, ATBv1.0 and ATBv1.1, ARM.

"The network simulator (ns-2)." www.isi.edu/nsnam/ns/.

"Noxim, the NoC Simulator."

"OPNET Technologies, Inc. OPNET Modeler." www.opnet.com.

"Simics."

"Spice."

"Virtual Socket Interface Alliance (VSIA)." Accessed 1st March 2016. http://vsi.org/

Agarwal, Ankur, Cyril Iskander, and Ravi Shankar. 2009. "Survey of network on chip (NoC) architectures & contributions." *Journal of Engineering, Computing and Architecture* 3 (1):21–27.

Agarwal, Niket, Tushar Krishna, Li-Shiuan Peh, and Niraj K Jha. 2009. "GARNET: A detailed on-chip network model inside a full-system simulator." *2009 IEEE International Symposium on Performance Analysis of Systems and Software.*

Ahmad, Balal, Ahmet T Erdogan, and Sami Khawam. 2006. "Architecture of a dynamically reconfigurable NoC for adaptive reconfigurable MPSoC." *First NASA/ESA Conference on Adaptive Hardware and Systems (AHS'06).*

Ahmed, Akram Ben, and Abderazek Ben Abdallah. 2014. "Graceful deadlock-free fault-tolerant routing algorithm for 3D Network-on-Chip architectures." *Journal of Parallel and Distributed Computing* 74 (4):2229–2240.

Ahmed, Akram Ben, and Abderazek Ben Abdallah. 2016. "Adaptive fault-tolerant architecture and routing algorithm for reliable many-core 3D-NoC systems." *Journal of Parallel and Distributed Computing* 93:30–43.

Ainsworth, Thomas William, and Timothy Mark Pinkston. 2007. "Characterizing the Cell EIB on-chip network." *IEEE Micro* 27 (5):6–14.

Al Faruque, Mohammad Abdullah, Thomas Ebi, and Jörg Henkel. 2010. "AdNoC: Runtime adaptive network-on-chip architecture." *IEEE Transactions on Very Large Scale Integration (VLSI) Systems* 20 (2):257–269.

Amde, Manish, Tomaz Felicijan, Aristeidis Efthymiou, Douglas Edwards, and Luciano Lavagno. 2005. "Asynchronous on-chip networks." *IEE Proceedings-Computers and Digital Techniques* 152 (2):273–283.

Anjo, Kenichiro, Yutaka Yamada, Michihiro Koibuchi, Akiya Jouraku, and Hideharu Amano. 2004. "BLACK-BUS: A new data-transfer technique using local address on networks-on-chips." *2004 18th International Parallel and Distributed Processing Symposium. Proceedings.*

ARM. " ARM." Accessed 1st October 2019. www.arm.com/products/system-ip/amba/amba-open-specifications.php

Arteris, SA. 2005. "A comparison of network-on-chip and busses." *White Paper.*

Bahirat, Shirish, and Sudeep Pasricha. 2014. "HELIX: Design and synthesis of hybrid nano-photonic application-specific network-on-chip architectures." *Fifteenth International Symposium on Quality Electronic Design.*

Bahrebar, Poona, and Dirk Stroobandt. 2014. "Adaptive and reconfigurable fault-tolerant routing method for 2D Networks-on-Chip." *2014 International Conference on ReConFigurable Computing and FPGAs (ReConFig14).*

Bainbridge, John, and Steve Furber. 2002. "Chain: A delay-insensitive chip area interconnect." *IEEE Micro* (5):16–23.

Bartic, TA, J-Y Mignolet, Vincent Nollet, Theodore Marescaux, Diederik Verkest, Serge Vernalde, and Rudy Lauwereins. 2005. "Topology adaptive network-on-chip design and implementation." *IEE Proceedings-Computers and Digital Techniques* 152 (4):467–472.

Beigné, Edith, Fabien Clermidy, Hélène Lhermet, Sylvain Miermont, Yvain Thonnart, Xuan-Tu Tran, Alexandre Valentian, Didier Varreau, Pascal Vivet, and Xavier Popon. 2009. "An asynchronous power aware and adaptive NoC based circuit." *IEEE Journal of Solid-State Circuits* 44 (4):1167–1177.

Beigné, Edith, Fabien Clermidy, Pascal Vivet, Alain Clouard, and Marc Renaudin. 2005. "An asynchronous NoC architecture providing low latency service and its multi-level design framework." *11th IEEE International Symposium on Asynchronous Circuits and Systems.*

Bell, Shane, Bruce Edwards, John Amann, Rich Conlin, Kevin Joyce, Vince Leung, John MacKay, Mike Reif, Liewei Bao, and John Brown. 2008. "Tile64-processor: A 64-core soc with mesh interconnect." *2008 IEEE International Solid-State Circuits Conference-Digest of Technical Papers.*

Benini, Luca, and Davide Bertozzi. 2005. "Network-on-chip architectures and design methods." *IEE Proceedings-Computers and Digital Techniques* 152 (2):261–272.

Benini, Luca, and Giovanni De Micheli. 2002. "Networks on chips: A new SoC paradigm." *Computer* 35 (1):70–78.

Ben-Itzhak, Yaniv, Eitan Zahavi, Israel Cidon, and Avinoam Kolodny. 2012. "HNOCS: Modular open-source simulator for heterogeneous NoCs." *2012 International Conference on Embedded Computer Systems (SAMOS).*

Bertozzi, Davide, and Luca Benini. 2004. "Xpipes: A network-on-chip architecture for gigascale systems-on-chip." *IEEE Circuits and Systems Magazine* 4 (2):18–31.

Bjerregaard, Tobias, and Shankar Mahadevan. 2006. "A survey of research and practices of network-on-chip." *ACM Computing Surveys (CSUR)* 38 (1):1.

Bjerregaard, Tobias, and Jens Sparsø. 2004. "Virtual channel designs for guaranteeing bandwidth in asynchronous network-on-chip." *2004 Proceedings Norchip Conference.*

Bjerregaard, Tobias, and Jens Sparsø. 2005a. "A router architecture for connection-oriented service guarantees in the MANGO clockless network-on-chip." *Design, Automation and Test in Europe.*

Bjerregaard, Tobias, and Jens Sparsø. 2005b. "A router architecture for connection-oriented service guarantees in the MANGO clockless network-on-chip." *2005 Design, Automation and Test in Europe. Proceedings.*

Bjerregaard, Tobias, and Jens Sparsø. 2005c. "Scheduling discipline for latency and bandwidth guarantees in asynchronous network-on-chip." *11th IEEE International Symposium on Asynchronous Circuits and Systems.*

Bobda, Christophe, Ali Ahmadinia, Mateusz Majer, Jürgen Teich, Sándor Fekete, and Jan van der Veen. 2005. "Dynoc: A dynamic infrastructure for communication in dynamically reconfigurable devices." *2005 International Conference on Field Programmable Logic and Applications.*

Bolotin, Evgeny, Israel Cidon, Ran Ginosar, and Avinoam Kolodny. 2004. "QNoC: QoS architecture and design process for network on chip." *Journal of Systems Architecture* 50 (2–3):105–128.

Bouhraoua, A, and ME Elrabaa. 2006. "A high-throughput network-on-chip architecture for systems-on-chip interconnect." *2006 International Symposium on System-on-Chip.*

Braun, Lars, Michael Hübner, Jürgen Becker, Thomas Perschke, Volker Schatz, and Stefan Bach. 2007. "Circuit switched run-time adaptive network-on-chip for image processing applications." *2007 International Conference on Field Programmable Logic and Applications.*

Brugge, Mike, and Mohammed AS Khalid. 2014. "A parameterizable NoC router for FPGAs." *JCP* 9 (3):519–528.

Butts, Mike. 2007. "Synchronization through communication in a massively parallel processor array." *IEEE Micro* 27 (5):32–40.

Castells-Rufas, David, Jaume Joven, and Jordi Carrabina. 2006. "A validation and performance evaluation tool for ProtoNoC." *2006 International Symposium on System-on-Chip.*

Castells-Rufas, David, Jaume Joven, Sergi Risueño, Eduard Fernandez, and Jordi Carrabina. 2009. "NocMaker: A cross-platform open-source design space exploration tool for networks on chip." *INA-OCMC Workshop*, Paphos, Cyprus.

Chan, Jeremy, and Sri Parameswaran. 2004. "NoCGEN: A template based reuse methodology for networks on chip architecture." *17th International Conference on VLSI Design. Proceedings.*

Chang, Kuei-Chung, Jih-Sheng Shen, and Tien-Fu Chen. 2006. "Evaluation and design trade-offs between circuit-switched and packet-switched NOCs for application-specific SOCs." *Proceedings of the 43rd Annual Design Automation Conference.*

Chi, Hsin-Chou, and Jia-Hung Chen. 2004. "Design and implementation of a routing switch for on-chip interconnection networks." *Proceedings of 2004 IEEE Asia-Pacific Conference on Advanced System Integrated Circuits.*

Chien, Andrew A, and Jae H Kim. 1992. *Planar-adaptive routing: Low-cost adaptive networks for multiprocessors*. Vol. 20: ACM.

Ching, Doris, Patrick Schaumont, and Ingrid Verbauwhede. 2004. "Integrated modeling and generation of a reconfigurable network-on-chip." *2004 18th International Parallel and Distributed Processing Symposium. Proceedings.*

Chiu, Ge-Ming. 2000. "The odd-even turn model for adaptive routing." *IEEE Transactions on Parallel and Distributed Systems* 11 (7):729–738.

Choudhary, Sudhanshu, and Shafi Qureshi. 2012. "Performance evaluation of mesh-based NoCs: Implementation of a new architecture and routing algorithm." *International Journal of Automation and Computing* 9 (4):403–413.

Coppola, Marcello, Riccardo Locatelli, Giuseppe Maruccia, Lorenzo Pieralisi, and Alberto Scandurra. 2004. "Spidergon: A novel on-chip communication network." *2004 International Symposium on System-on-Chip. Proceedings.*

Dally, William J, and Brian Towles. 2001. "Route packets, not wires: On-chip interconnection networks." *Proceedings of the 38th Annual Design Automation Conference.*

Dielissen, John, Andrei Rădulescu, Kees Goossens, and Edwin Rijpkema. 2003. "Concepts and implementation of the Philips network-on-chip." *IP-Based SoC Design.*

DiTomaso, Dominic, Avinash Kodi, and Ahmed Louri. 2014. "QORE: A fault tolerant network-on-chip architecture with power-efficient quad-function channel (QFC) buffers." *2014 IEEE 20th International Symposium on High Performance Computer Architecture (HPCA).*

Dobkin, Rostislav, Ran Ginosar, and Israel Cidon. 2007. "QNoC asynchronous router with dynamic virtual channel allocation." *First International Symposium on Networks-on-Chip (NOCS'07)*.

El-Moursy, Magdy A, Darek Korzec, and Mohammed Ismail. 2009. "High throughput architecture for OCTAGON network on chip." *2009 16th IEEE International Conference on Electronics, Circuits and Systems-(ICECS 2009)*.

Emerson, K. 1997. "Asynchronous design-An interesting alternative." *Proceedings Tenth International Conference on VLSI Design*.

Evain, Samuel, J-P Diguet, and Dominique Houzet. 2004. "μ spider: A CAD tool for efficient NoC design." *2004 Proceedings Norchip Conference*.

Feero, Brett Stanley, and Partha Pratim Pande. 2008. "Networks-on-chip in a three-dimensional environment: A performance evaluation." *IEEE Transactions on Computers* 58 (1):32–45.

Feliciian, F, and Stephen B Furber. 2004. "An asynchronous on-chip network router with quality-of-service (QoS) support." *2004 IEEE International SOC Conference. Proceedings*.

Feng, Chaochao, Zhonghai Lu, Axel Jantsch, Minxuan Zhang, and Zuocheng Xing. 2012. "Addressing transient and permanent faults in NoC with efficient fault-tolerant deflection router." *IEEE Transactions on Very Large Scale Integration (VLSI) Systems* 21 (6):1053–1066.

Forsell, Martti. 2002. "A scalable high-performance computing solution for networks on chips." *IEEE Micro* 22 (5):46–55.

Ganguly, Amlan, Kevin Chang, Sujay Deb, Partha Pratim Pande, Benjamin Belzer, and Christof Teuscher. 2010. "Scalable hybrid wireless network-on-chip architectures for multicore systems." *IEEE Transactions on Computers* 60 (10):1485–1502.

Glass, Christopher J, and Lionel M Ni. 1992. "The turn model for adaptive routing." *ACM SIGARCH Computer Architecture News* 20 (2):278–287.

Göhringer, Diana, Michael Hübner, Laure Hugot-Derville, and Jürgen Becker. 2010. "Message passing interface support for the runtime adaptive multi-processor system-on-chip RAMPSoC." *2010 International Conference on Embedded Computer Systems: Architectures, Modeling and Simulation*.

Goossens, Kees, John Dielissen, and Andrei Rădulescu. 2005. "Æthereal network on chip: Concepts, architectures, and implementations." *IEEE Design & Test of Computers* 22 (5):414–421.

Gratz, Paul, Changkyu Kim, Karthikeyan Sankaralingam, Heather Hanson, Premkishore Shivakumar, Stephen W Keckler, and Doug Burger. 2007. "On-chip interconnection networks of the TRIPS chip." *IEEE Micro* 27 (5):41–50.

Grot, Boris, Joel Hestness, Stephen W Keckler, and Onur Mutlu. 2011. "Kilo-NoC: A heterogeneous network-on-chip architecture for scalability and service guarantees." *ACM SIGARCH Computer Architecture News* 39 (3):401–412.

Guerrier, Pierre, and Alain Greiner. 2008. "A generic architecture for on-chip packet-switched interconnections." *Design, Automation, and Test in Europe*.

Hansson, Andreas, Kees Goossens, and Andrei Rădulescu. 2007. "Avoiding message-dependent deadlock in network-based systems on chip." *VLSI Design 2007*.

Hansson, Andreas, Mahesh Subburaman, and Kees Goossens. 2009. "Aelite: A flit-synchronous network on chip with composable and predictable services." *Proceedings of the Conference on Design, Automation and Test in Europe*.

Heisswolf, Jan, Andreas Weichslgartner, Aurang Zaib, Stephanie Friederich, Leonard Masing, Carsten Stein, Marco Duden, Roman Klöpfer, Jürgen Teich, and Thomas Wild. 2015. "Fault-tolerant communication in invasive networks on chip." *2015 NASA/ESA Conference on Adaptive Hardware and Systems (AHS)*.

Henriksson, Tomas, Daniel Wiklund, and Dake Liu. 2003. "VLSI implementation of a switch for on-chip networks." *Proceedings of the International Workshop on Design and Diagnostics of Electronic Circuits and Systems*, Poznan, Poland.

Hilton, Clint, and Brent Nelson. 2006. "PNoC: A flexible circuit-switched NoC for FPGA-based systems." *IEE Proceedings-Computers and Digital Techniques* 153 (3):181–188.

Hoffman, Jeff, David Arditti Ilitzky, Anthony Chun, and Aliaksei Chapyzhenka. 2007. "Architecture of the scalable communications core." *First International Symposium on Networks-on-Chip (NOCS'07)*.

Hosseinabady, Mohammad, Mohammad Reza Kakoee, Jimson Mathew, and Dhiraj K Pradhan. 2007. "Reliable network-on-chip based on generalized de Bruijn graph." *2007 IEEE International High Level Design Validation and Test Workshop*.

Hu, Jingcao, and Radu Marculescu. 2003. "Exploiting the routing flexibility for energy/performance aware mapping of regular NoC architectures." *2003 Design, Automation and Test in Europe Conference and Exhibition*.

Hu, Jingcao, and Radu Marculescu. 2004. "DyAD: Smart routing for networks-on-chip." *Proceedings of the 41st Annual Design Automation Conference*.

Jain, Lavina, B Al-Hashimi, MS Gaur, V Laxmi, and A Narayanan. 2007. "NIRGAM: A simulator for NoC interconnect routing and application modeling." *Design, Automation and Test in Europe Conference*.

Janarthanan, Arun, Vijay Swaminathan, and Karen A Tomko. 2007. "MoCReS: An area-efficient multi-clock on-chip network for reconfigurable systems." *IEEE Computer Society Annual Symposium on VLSI (ISVLSI'07)*.

Janarthanan, Arun, and Karen A Tomko. 2008. "MoCSYS: A multi-clock hybrid two-layer router architecture and integrated topology synthesis framework for system-level design of FPGA based on-chip networks." *21st International Conference on VLSI Design (VLSID 2008)*.

Kao, Yu-Hsiang, and H Jonathan Chao. 2012. "Design of a bufferless photonic clos network-on-chip architecture." *IEEE Transactions on Computers* 63 (3):764–776.

Kao, Yu-Hsiang, and H Jonathan Chao. 2014. "Design of a bufferless photonic clos network-on-chip architecture." *IEEE Transactions on Computers* 63 (3):764–776.

Karim, Faraydon, Anh Nguyen, and Sujit Dey. 2002. "An interconnect architecture for networking systems on chips." *IEEE Micro* 22 (5):36–45.

Kariniemi, Heikki, and Jari Nurmi. 2006. "On-line reconfigurable XGFT network-on-chip designed for improving the fault-tolerance and manufacturability of the MPSoC chips." *2006 International Conference on Field Programmable Logic and Applications*.

Kavaldjiev, Nikolay, Gerard JM Smit, Pierre G Jansen, and Pascal T Wolkotte. 2006. "A virtual channel network-on-chip for GT and BE traffic." *IEEE Computer Society Annual Symposium on Emerging VLSI Technologies and Architectures (ISVLSI'06)*.

Kavaldjiev, Nikolay, Gerard JM Smit, Pascal T Wolkotte, and Pierre G Jansen. 2006. "Providing QoS guarantees in a NoC by virtual channel reservation." *International Workshop on Applied Reconfigurable Computing*.

Kia, Hamed Sajjadi, Cristinel Ababei, Sudarshan Srinivasan, and Shaista Jabeen. 2015. "A new scalable fault tolerant routing algorithm for networks-on-chip." *2015 IEEE 58th International Midwest Symposium on Circuits and Systems (MWSCAS)*.

Kim, Daewook, Manho Kim, and Gerald E Sobelman. 2005. "Design of a high-performance scalable CDMA router for on-chip switched networks." 대한전자공학회 *ISOCC*:32–35.

Kim, Jongman, Dongkook Park, Theo Theocharides, Narayanan Vijaykrishnan, and Chita R Das. 2005. "A low latency router supporting adaptivity for on-chip interconnects." *2005 Proceedings. 42nd Design Automation Conference*.

Kim, Young Bok, and Yong-Bin Kim. 2007. "Fault tolerant source routing for network-on-chip." *22nd IEEE International Symposium on Defect and Fault-Tolerance in VLSI Systems (DFT 2007)*.

Krasteva, Yana E, Eduardo De la Torre, and Teresa Riesgo. 2010. "Reconfigurable networks on chip: DRNoC architecture." *Journal of Systems Architecture* 56 (7):293–302.

Kumar, Amit, Li-Shiuan Peh, Partha Kundu, and Niraj K Jha. 2008. "Toward ideal on-chip communication using express virtual channels." *IEEE Micro* 28 (1):80–90.

Kumar, Shashi, Axel Jantsch, J-P Soininen, Martti Forsell, Mikael Millberg, Johny Oberg, Kari Tiensyrja, and Ahmed Hemani. 2002. "A network on chip architecture and design methodology." *Proceedings IEEE Computer Society Annual Symposium on VLSI. New Paradigms for VLSI Systems Design (ISVLSI 2002).*

Lan, Ying-Cherng, Hsiao-An Lin, Shih-Hsin Lo, Yu Hen Hu, and Sao-Jie Chen. 2011. "A bidirectional NoC (BiNoC) architecture with dynamic self-reconfigurable channel." *IEEE Transactions on Computer-Aided Design of Integrated Circuits and Systems* 30 (3):427–440.

Lattard, Didier, Edith Beigné, Fabien Clermidy, Yves Durand, Romain Lemaire, Pascal Vivet, and Friedbert Berens. 2008. "A reconfigurable baseband platform based on an asynchronous network-on-chip." *IEEE Journal of Solid-State Circuits* 43 (1):223–235.

Le Beux, Sébastien, Jelena Trajkovic, Ian O'Connor, Gabriela Nicolescu, Guy Bois, and Pierre Paulin. 2011. "Optical ring network-on-chip (ORNoC): Architecture and design methodology." *2011 Design, Automation & Test in Europe.*

Lee, Hyung Gyu, Umit Y Ogras, Radu Marculescu, and Naehyuck Chang. 2006. "Design space exploration and prototyping for on-chip multimedia applications." *Proceedings of the 43rd Annual Design Automation Conference.*

Lee, Junghee, Chrysostomos Nicopoulos, Sung Joo Park, Madhavan Swaminathan, and Jongman Kim. 2013. "Do we need wide flits in networks-on-chip?" *2013 IEEE Computer Society Annual Symposium on VLSI (ISVLSI).*

Lee, Kangmin, Se-Joong Lee, Sung-Eun Kim, Hye-Mi Choi, Donghyun Kim, Sunyoung Kim, Min-Wuk Lee, and Hoi-Jun Yoo. 2004. "A 51mW 1.6 GHz on-chip network for low-power heterogeneous SoC platform." *2004 IEEE International Solid-State Circuits Conference (IEEE Cat. No. 04CH37519).*

Lee, Kangmin, Se-Joong Lee, and Hoi-Jun Yoo. 2006. "Low-power network-on-chip for high-performance SoC design." *IEEE Transactions on Very Large Scale Integration (VLSI) Systems* 14 (2):148–160.

Lee, Sanghun, Chanho Lee, and Hyuk-Jae Lee. 2004. "A new multi-channel on-chip-bus architecture for system-on-chips." *2004 IEEE International SOC Conference. Proceedings.*

Lee, Se-Joong, Kwanho Kim, Hyejung Kim, Namjun Cho, and Hoi-Jun Yoo. 2005. "Adaptive network-on-chip with wave-front train serialization scheme." *2005 Digest of Technical Papers. 2005 Symposium on VLSI Circuits.*

Lee, Se-Joong, Seong-Jun Song, Kangmin Lee, Jeong-Ho Woo, Sung-Eun Kim, Byeong-Gyu Nam, and Hoi-Jun Yoo. 2003. "An 800MHz star-connected on-chip network for application to systems on a chip." *2003 IEEE International Solid-State Circuits Conference. Digest of Technical Papers. ISSCC.*

Lee, Seungju, Nozomu Togawa, Yusuke Sekihara, Takashi Aoki, and Akira Onozawa. 2012. "A hybrid NoC architecture utilizing packet transmission priority control method." *2012 IEEE Asia Pacific Conference on Circuits and Systems.*

Leroy, Anthony, Paul Marchal, Adelina Shickova, Francky Catthoor, Frédéric Robert, and Diederik Verkest. 2005. "Spatial division multiplexing: A novel approach for guaranteed throughput on NoCs." *Proceedings of the 3rd IEEE/ACM/IFIP International Conference on Hardware/Software Codesign and System Synthesis.*

Liang, Jian, Andrew Laffely, Sriram Srinivasan, and Russell Tessier. 2004. "An architecture and compiler for scalable on-chip communication." *IEEE Transactions on Very Large Scale Integration (VLSI) Systems* 12 (7):711–726.

Liang, Jian, Sriram Swaminathan, and Russell Tessier. 2000. "aSOC: A scalable, single-chip communications architecture." *Proceedings 2000 International Conference on Parallel Architectures and Compilation Techniques (Cat. No. PR00622).*

Lines, Andrew. 2004. "Asynchronous interconnect for synchronous SoC design." *IEEE Micro* 24 (1):32–41.

Logvinenko, Alexander, Carsten Gremzow, and Dietmar Tutsch. 2013. "RecMIN: A reconfiguration architecture for network on chip." *2013 8th International Workshop on Reconfigurable and Communication-Centric Systems-on-Chip (ReCoSoC).*

Logvinenko, Alexander, and Dietmar Tutsch. 2012. "RECSIM-A simulator for reconfigurable network on chip topologies." *Proceedings of the 26th European Simulation and Modelling Conference (ESM 2012).* Essen, Germany.

Marculescu, Radu, Umit Y Ogras, Li-Shiuan Peh, Natalie Enright Jerger, and Yatin Hoskote. 2008. "Outstanding research problems in NoC design: System, microarchitecture, and circuit perspectives." *IEEE Transactions on Computer-Aided Design of Integrated Circuits and Systems* 28 (1):3–21.

Millberg, Mikael, Erland Nilsson, Rikard Thid, and Axel Jantsch. 2004. "Guaranteed bandwidth using looped containers in temporally disjoint networks within the Nostrum network on chip." *Proceedings Design, Automation and Test in Europe Conference and Exhibition.*

Mishra, Prabhakar, A Nidhi, and JK Kishore. 2012. "Custom Network on Chip architecture for map generation in autonomous navigating robots." *2012 Annual IEEE India Conference (INDICON).*

Mitić, Milica, and Mile Stojčev. 2006a. "An overview of on-chip buses." *Facta universitatis-series: Electronics and Energetics* 19 (3):405–428.

Mitić, Milica, and Mile Stojčev. 2006b. "A survey of three system-on-chip buses: Amba, coreconnect and wishbone." *Proceeding of the 41st International Conference on Information Communication Energy Systems and Technology (ICEST).*

Mondinelli, Filippo, Michele Borgatti, and Zsolt M KOVACS VAJNA. 2004. "A 0.13 um 1Gb/s/channel store-and-forward network on-chip." *IEEE Systems-on-Chip Conference.*

Moraes, Fernando, Ney Calazans, Aline Mello, Leandro Möller, and Luciano Ost. 2004. "HERMES: An infrastructure for low area overhead packet-switching networks on chip." *INTEGRATION, the VLSI Journal* 38 (1):69–93.

Moriam, Sadia, and Gerhard P Fettweis. 2016. "Fault tolerant deadlock-free adaptive routing algorithms for hexagonal networks-on-chip." *2016 EUROMICRO Conference on Digital System Design (DSD).*

Mullins, Robert, Andrew West, and Simon Moore. 2006. "The design and implementation of a low-latency on-chip network." *Proceedings of the 2006 Asia and South Pacific Design Automation Conference.*

Muralimanohar, Naveen, Rajeev Balasubramonian, and Norm Jouppi. 2007. "Optimizing NUCA organizations and wiring alternatives for large caches with CACTI 6.0." *Proceedings of the 40th Annual IEEE/ACM International Symposium on Microarchitecture.*

Neeb, Christian, and Norbert Wehn. 2008. "Designing efficient irregular networks for heterogeneous systems-on-chip." *Journal of Systems Architecture* 54 (3–4):384–396.

Nunez-Yanez, José L, Doug Edwards, and Antonio Marcello Coppola. 2008. "Adaptive routing strategies for fault-tolerant on-chip networks in dynamically reconfigurable systems." *IET Computers & Digital Techniques* 2 (3):184–198.

Nurmi, Jari. 2005. "Network-on-chip: A new paradigm for system-on-chip design." *2005 International Symposium on System-on-Chip.*

Ogras, Umit Y, Jingcao Hu, and Radu Marculescu. 2005. "Key research problems in NoC design: A holistic perspective." *Proceedings of the 3rd IEEE/ACM/IFIP International Conference on Hardware/software Codesign and System Synthesis.*

Ogras, Umit Y, Radu Marculescu, Hyung Gyu Lee, and Naehyuck Chang. 2006. "Communication architecture optimization: Making the shortest path shorter in regular networks-on-chip." *Proceedings of the Conference on Design, Automation and Test in Europe.*

Palermo, Gianluca, Cristina Silvano, Giovanni Mariani, Riccardo Locatelli, and Marcello Coppola. 2007. "Application-specific topology design customization for STNoC." *10th EUROMICRO Conference on Digital System Design Architectures, Methods and Tools (DSD 2007).*

Palesi, Maurizio, and Masoud Daneshtalab. 2014. *Routing algorithms in networks-on-chip*: Springer.

Panades, Ivan Miro, Alain Greiner, and Abbas Sheibanyrad. 2006. "A low cost network-on-chip with guaranteed service well suited to the GALS approach." *2006 1st International Conference on Nano-Networks and Workshops*.

Pande, Partha Pratim, Cristian Grecu, André Ivanov, and Res Saleh. 2003. "High-throughput switch-based interconnect for future SoCs." *2003 3rd IEEE International Workshop on System-on-Chip for Real-Time Applications. Proceedings*.

Pande, Partha Pratim, Cristian Grecu, Michael Jones, André Ivanov, and Resve Saleh. 2005. "Performance evaluation and design trade-offs for network-on-chip interconnect architectures." *IEEE Transactions on Computers* 54 (8):1025–1040.

Pasricha, Sudeep, and Yong Zou. 2011. "A low overhead fault tolerant routing scheme for 3D Networks-on-Chip." *2011 12th International Symposium on Quality Electronic Design*.

Patooghy, Ahmad, and Seyed Ghassem Miremadi. 2010. "Complement routing: A methodology to design reliable routing algorithm for Network on Chips." *Microprocessors and Microsystems* 34 (6):163–173.

Paukovits, Christian, and Hermann Kopetz. 2008. "Concepts of switching in the time-triggered network-on-chip." *2008 14th IEEE International Conference on Embedded and Real-Time Computing Systems and Applications*.

Peña-Ramos, JC, and Ramon Parra-Michel. 2011. "Network on chip architectures for high performance digital signal processing using a configurable core." *2011 International Conference on Reconfigurable Computing and FPGAs*.

Penolazzi, Sandro, and Axel Jantsch. 2006. "A high level power model for the Nostrum NoC." *9th EUROMICRO Conference on Digital System Design (DSD'06)*.

Pionteck, Thilo, Roman Koch, and Carsten Albrecht. 2006. "Applying partial reconfiguration to networks-on-chips." *2006 International Conference on Field Programmable Logic and Applications*.

Pirretti, Matthew, Greg M Link, Richard R Brooks, Narayanan Vijaykrishnan, Mahmut Kandemir, and Mary Jane Irwin. 2004. "Fault tolerant algorithms for network-on-chip interconnect." *IEEE Computer Society Annual Symposium on VLSI*.

Poluri, Pavan, and Ahmed Louri. 2014. "A soft error tolerant network-on-chip router pipeline for multi-core systems." *IEEE Computer Architecture Letters* 14 (2):107–110.

Postman, Jacob, Tushar Krishna, Christopher Edmonds, Li-Shiuan Peh, and Patrick Chiang. 2012. "Swift: A low-power network-on-chip implementing the token flow control router architecture with swing-reduced interconnects." *IEEE Transactions on Very Large Scale Integration (VLSI) Systems* 21 (8):1432–1446.

Qouneh, Amer, Zhongqi Li, Madhura Joshi, Wangyuan Zhang, Xin Fu, and Tao Li. 2012. "Aurora: A thermally resilient photonic network-on-chip architecture." *2012 IEEE 30th International Conference on Computer Design (ICCD)*.

Rădulescu, Andrei, and Kees Goossens. 2004. "Communication services for networks on chip." *Domain-Specific Processors: Systems, Architectures, Modeling, and Simulation*:193–213.

Rantala, Ville, Teijo Lehtonen, and Juha Plosila. 2006. *Network on chip routing algorithms*: Citeseer.

Rijpkema, Edwin, Kees Goossens, Andrei Rădulescu, John Dielissen, Jef van Meerbergen, Paul Wielage, and Erwin Waterlander. 2003. "Trade-offs in the design of a router with both guaranteed and best-effort services for networks on chip." *IEE Proceedings-Computers and Digital Techniques* 150 (5):294–302.

Sabbaghi-Nadooshan, Reza, Mehdi Modarressi, and Hamid Sarbazi-Azad. 2010. "A Novel De Bruijn based mesh topology for networks-on-chip." *VLSI*.

Salminen, Erno, Tero Kangas, Timo D Hämäläinen, Jouni Riihimäki, Vesa Lahtinen, and Kimmo Kuusilinna. 2006. "HIBI communication network for system-on-chip." *Journal of VLSI Signal Processing Systems for Signal, Image and Video Technology* 43 (2–3):185–205.

Samman, Faizal A, Thomas Hollstein, and Manfred Glesner. 2009. "Networks-on-chip based on dynamic wormhole packet identity mapping management." *VLSI Design*:2.

Samuelsson, Henrik, and Shashi Kumar. 2004. "Ring road NoC architecture." *2004 Proceedings Norchip Conference*.

Sanusi, Azeez, and Magdy A Bayoumi Nan Wang. 2006. "A central caching network-on-chip communication architecture design." https://www.design-reuse.com/articles/15634/a-central-caching-network-on-chip-communication-architecture-design.html

Schoeberl, Martin. 2007. "A time-triggered network-on-chip." *2007 International Conference on Field Programmable Logic and Applications*.

Schonwald, Timo, Oliver Bringmann, and Wolfgang Rosenstiel. 2007. "Region-based routing algorithm for network-on-chip architectures." *Norchip 2007*.

Schonwald, Timo, Jochen Zimmermann, Oliver Bringmann, and Wolfgang Rosenstiel. 2007. "Fully adaptive fault-tolerant routing algorithm for network-on-chip architectures." *10th EUROMICRO Conference on Digital System Design Architectures, Methods and Tools (DSD 2007)*.

Schuck, Christian, Stefan Lamparth, and Jürgen Becker. 2007. "artNoC-A novel multi-functional router architecture for Organic Computing." *2007 International Conference on Field Programmable Logic and Applications*.

Seifi, Mohammad Reza, and Mohammad Eshghi. 2012. "Clustered NoC, a suitable design for group communications in Network on Chip." *Computers & Electrical Engineering* 38 (1):82–95.

Shu, Hao, Jiang-Yi Shi, Yue Hao, Pei-Jun Ma, and Zhao Xu. 2012. "DANoC: A dynamic adaptive network on chip architecture." *2012 IEEE 11th International Conference on Solid-State and Integrated Circuit Technology*.

Sigüenza-Tortosa, David, Tapani Ahonen, and Jari Nurmi. 2004. "Issues in the development of a practical NoC: The Proteo concept." *Integration, the VLSI Journal* 38 (1):95–105.

Sigüenza-Tortosa, David, and Jari Nurmi. 2002. "Proteo: A new approach to network-on-chip." *Proceedings, IASTED-Communication Systems and Networks (CSN 2002)*.

Song, Wei, and Doug Edwards. 2011. "Asynchronous spatial division multiplexing router." *Microprocessors and Microsystems* 35 (2):85–97.

Soteriou, Vassos, Noel Eisley, Hangsheng Wang, Bin Li, and Li-Shiuan Peh. 2006. "Polaris: A system-level roadmap for on-chip interconnection networks." *2006 International Conference on Computer Design*.

Specification, AMBA. 1999. "Rev. 2.0," *ARM*. www. arm. com.

Stefan, Radu Andrei, Anca Molnos, and Kees Goossens. 2012. "dAElite: A TDM NoC supporting QoS, multicast, and fast connection set-up." *IEEE Transactions on Computers* 63 (3):583–594.

Stensgaard, Mikkel Bystrup, and Jens Sparsø. 2008. "ReNoC: A network-on-chip architecture with reconfigurable topology." *Proceedings of the Second ACM/IEEE International Symposium on Networks-on-Chip*.

Tatas, Konstantinos, Kostas Siozios, Dimitrios Soudris, and Axel Jantsch. 2014. *Designing 2D and 3D network-on-chip architectures*: Springer.

Taylor, Michael Bedford, Jason Kim, Jason Miller, David Wentzlaff, Fae Ghodrat, Ben Greenwald, Henry Hoffman, Paul Johnson, Jae-Wook Lee, and Walter Lee. 2002. "The raw microprocessor: A computational fabric for software circuits and general-purpose programs." *IEEE Micro* 22 (2):25–35.

Teehan, Paul, Mark Greenstreet, and Guy Lemieux. 2007. "A survey and taxonomy of GALS design styles." *IEEE Design & Test of Computers* 24 (5):418–428.

Vangal, Sriram R, Jason Howard, Gregory Ruhl, Saurabh Dighe, Howard Wilson, James Tschanz, David Finan, Arvind Singh, Tiju Jacob, and Shailendra Jain. 2008. "An 80-tile sub-100-w teraflops processor in 65-nm CMOS." *IEEE Journal of Solid-State Circuits* 43 (1):29–41.

Varga, András, and Rudolf Hornig. 2008. "An overview of the OMNeT++ simulation environment." *Proceedings of the 1st International Conference on Simulation Tools and Techniques for Communications, Networks and Systems & Workshops.*

Wang, Chifeng, and Nader Bagherzadeh. 2014. "Design and evaluation of a high throughput QoS-aware and congestion-aware router architecture for network-on-chip." *Microprocessors and Microsystems* 38 (4):304–315.

Wang, Chifeng, Wen-Hsiang Hu, Seung Eun Lee, and Nader Bagherzadeh. 2011. "Area and power-efficient innovative congestion-aware Network-on-Chip architecture." *Journal of Systems Architecture* 57 (1):24–38.

Wang, Hang-Sheng, Xinping Zhu, Li-Shiuan Peh, and Sharad Malik. 2002. "Orion: A power-performance simulator for interconnection networks." *Proceedings of the 35th Annual ACM/IEEE International Symposium on Microarchitecture.*

Wang, Nan, Azeez Sanusi, PY Zhao, M Elgamel, and Magdy A Bayoumi. 2010. "PMCNOC: A pipelining multi-channel central caching network-on-chip communication architecture design." *Journal of Signal Processing Systems* 60 (3):315–331.

Wang, Xin, Tapani Ahonen, and Jari Nurmi. 2007. "Applying CDMA technique to network-on-chip." *IEEE Transactions on Very Large Scale Integration (VLSI) Systems* 15 (10):1091–1100.

Wiklund, Daniel, and Dake Liu. 2003. "SoCBUS: Switched network on chip for hard real time embedded systems." *Proceedings International Parallel and Distributed Processing Symposium.*

Wingard, Drew. 2001. "MicroNetwork-based integration for SOCs." *Proceedings of the 38th Design Automation Conference (IEEE Cat. No. 01CH37232).*

Wolkotte, Pascal T, Gerard JM Smit, Gerard K Rauwerda, and Lodewijk T Smit. 2005. "An energy-efficient reconfigurable circuit-switched network-on-chip." *19th IEEE International Parallel and Distributed Processing Symposium.*

Wu, Jie. 2000. "A fault-tolerant adaptive and minimal routing approach in nD meshes." *Proceedings 2000 International Conference on Parallel Processing.*

Wu, Li-Wei, Wei-Xiang Tang, and Yarsun Hsu. 2011. "A novel architecture and routing algorithm for dynamic reconfigurable network-on-chip." *2011 IEEE Ninth International Symposium on Parallel and Distributed Processing with Applications.*

Yaghini, Pooria M, Ashkan Eghbal, SA Asghari, and H Pedram. 2009. "Power comparison of an asynchronous and synchronous network on chip router." *2009 14th International CSI Computer Conference.*

Yaghini, Pooria M, Ashkan Eghbal, and Nader Bagherzadeh. 2015. "On the design of hybrid routing mechanism for mesh-based network-on-chip." *Integration* 50:183–192.

Yang, Yoon Seok, Reeshav Kumar, Gwan Choi, and Paul V Gratz. 2014. "WaveSync: Low-latency source-synchronous bypass network-on-chip architecture." *ACM Transactions on Design Automation of Electronic Systems (TODAES)* 19 (4):34.

Zeferino, Cesar Albenes, Márcio Eduardo Kreutz, and Altamiro Amadeu Susin. 2004. "RASoC: A router soft-core for networks-on-chip." *Proceedings of the Conference on Design, Automation and Test in Europe-Volume 3.*

Zeferino, Cesar Albenes, and Altamiro Amadeu Susin. 2003. "SoCIN: A parametric and scalable network-on-chip." *2003 16th Symposium on Integrated Circuits and Systems Design (SBCCI 2003). Proceedings.*

Zhu, Haibo, Partha Pratim Pande, and Cristian Grecu. 2007. "Performance evaluation of adaptive routing algorithms for achieving fault tolerance in NoC fabrics." *2007 IEEE International Conf. on Application-specific Systems, Architectures and Processors (ASAP).*

Zid, Mounir, Abdelkrim Zitouni, Adel Baganne, and Rached Tourki. 2006. "New generic GALS NoC architectures with multiple QoS." *2006 International Conference on Design and Test of Integrated Systems in Nanoscale Technology (DTIS 2006).*

3 Bio-Inspired Algorithms and Implementation

3.1 SWARM INTELLIGENCE ALGORITHMS

Swarm intelligence algorithms (SIAs) are inspired by the behavior of social animals like ants, birds, fish and termites. Particle swarm optimization (PSO) is an example of SIA. The PSO algorithm is inspired by the movement of flocks of birds and schools of fish. PSO is being used in computational science to solve complex problems by repeatedly trying to improve the particular solution from several possible particles (solutions) (Parpinelli and Lopes 2011).

3.2 ANT COLONY OPTIMIZATION

The ant colony optimization (ACO) algorithm is based on the foraging behavior of ants. ACO has been adopted in the organization of parallel distributed systems. In the biological world, ants collaborate with each other and release a special liquid (called a pheromone) when they find food. Later, ants group together and follow this shortest path constructed because of the release of this liquid (Drigo 1996).

3.3 ARTIFICIAL IMMUNE SYSTEM

The artificial immune system (AIS) is inspired by the biological immune system. AIS is used to detect the environmental changes from the typical system and memorize these characteristics automatically (Castro, De Castro, and Timmis 2002).

3.4 FIREFLY ALGORITHM

The flashing pattern of a firefly acts as a communication pattern between fireflies. Fireflies also use these flashing patterns to attract prey. In addition, the fireflies are attracted to each other by the intensity of these flashing lights that affect the distance between them. The optimization problems are inspired by the firefly algorithms through which problems are being solved with the brightness of the optimal solution (Yang 2010a).

3.5 EPIDEMIC SPREADING

The dissemination of information in wireless *ad hoc* networks is inspired by the epidemic spreading technique. The distribution of information particles or the spread of viruses on the Web is linked with the dissemination of information (Vogels, Van Renesse, and Birman 2003).

3.6 FLOWER POLLINATION ALGORITHM

The flower pollination algorithm (FPA) got its inspiration from the natural pollination process. Self- and cross-pollination take place using biotic and abiotic mechanisms. Animals and insects use biotic pollination, while water and wind help in abiotic pollination. A flower or pollen in FPA is a possible solution to the optimization problem (Yang 2012).

3.7 ARTIFICIAL BEE COLONY ALGORITHM

There are three types of bees in the artificial bee colony (ABC) algorithm; employed bees, unemployed bees and onlooker bees. For every food source, the bees are employed. The employed bees go to the source of the food and later come back to the hive and dance. The bees select a food source based on its closeness to the hive, taste and richness of the energy. The bees become unemployed when the food source is finished; later these bees start searching for more sources of food. The onlooker bees select a food source based on the dance of the employed bees. In an ABC-inspired algorithm, the position of the food source becomes a possible solution to the optimization problem. Moreover, the amount of the food reflects the quality of the solution. The number of solutions in any problem depends upon the number of employed bees (Karaboga and Akay 2009).

3.8 CAT SWARM OPTIMIZATION

In the cat swarm optimization (CSO) algorithm, the behavior of cats is used to solve the problems of optimization. The solution set depends on the seeking and tracing modes of a cat. The seeking mode of the CSO algorithm is inspired by the behavior of a cat at rest and thinking about a new area to move to and change its current position based on the best possible candidate solutions. The tracing mode of CSO is inspired by the behavior of the cat when it is following a particular target. In the tracing mode, the cat tries to select a solution that has the best fitness value (Chu, Tsai, and Pan 2006).

3.9 CUCKOO SEARCH

The cuckoo search (CS) algorithm is used to solve complex engineering optimization problems. The CS algorithm is inspired by the brood parasitism mechanism of cuckoo birds. The solution obtained from the CS algorithm is the best among the different optimization algorithms due to the extensive search for the best solution. In this bio-inspired algorithm, a cuckoo lays their eggs in the nest of the other host species' nest. At times, the host birds throw out the cuckoo's egg, while some host birds leave their old nest and start building a new nest somewhere else. *Tapera*, a specialized cuckoo species, have evolved using the same mechanism (Yang and Deb 2009).

Bio-Inspired Algorithms and Implementation

3.10 BAT ALGORITHM

The bat algorithm (BA) is inspired by the echolocation ability of bats. The best solution is chosen from a number of possible solutions based on the optimal frequency, velocity and location of the prey. The bat emits a very high sound signal and listens for the echo that comes back from different surrounding objects. The echo received by the bat depends upon the type and location of the prey (Yang 2010b).

3.11 CUTTLEFISH ALGORITHM

The cuttlefish algorithm (CFA) is a search algorithm inspired by the color-changing mechanism of cuttlefish. Reflection and visibility are two strategies used in CFA to find the optimal solution (Eesa, Brifcani, and Orman 2013).

3.12 HARRIS HAWKS OPTIMIZATION

Harris hawks optimization (HHO) is inspired by the hunting style of Harris hawks. Hawks attack the prey from different directions to surprise it. Hawks also adapt different strategies to grasp the prey under different conditions. In HHO, different strategies are used to solve and optimize a complex problem (Heidari et al. 2019).

3.13 KILLER WHALE ALGORITHM

The killer whale algorithm (KWA) is inspired by the killing strategy of these creatures. The leader of the killer whale pod searches for the optimum position of the prey and the best direction to attack. The members of the killer whale pod are responsible for tracking and following the prey. KWA solves and finds the best solution based on the search and chase strategy of the killer whale. The memorizing capability of the killer whale is also incorporated in the KWA (Biyanto et al. 2017).

3.14 COBWEB NETWORK ON CHIP TOPOLOGY

The cobweb topology is inspired by the web made by a spider. The cobweb topology performs better compared to traditional mesh and torus topologies in terms of cost, latency and throughput. Cobweb topology is scalable and can be extended to any Network on Chip (NoC) size (Mishra, Nidhi, and Kishore 2014).

3.15 SCALABLE BIO-INSPIRED FAULT DETECTION UNIT IN NETWORK ON CHIP

The novel bio-inspired fault detection unit (FDU) is inspired by the synapse mechanism of a biological brain. The fault is detected using excitatory and inhibitory responses of synapses. The router is incorporated with FDU to inform the routing algorithm about faults. FDU also helps in the recovery of faulty messages in the NoC (McElholm et al. 2018).

3.16 AUTONOMOUS ERROR TOLERANT ARCHITECTURE

The autonomous error tolerant (AET) architecture is inspired by the biological brain's plasticity mechanism. Those neuron cells and synapses that die can be recovered by other neurons and synapses, which are active and fire at a high rate. Similarly, error cells (routers and interconnects) can be bypassed by active cells, and a single faulty cell does not affect the overall system's performance (Liu et al. 2017).

3.17 SPINNAKER COMMUNICATION

SpiNNaker is a multiple-chip system designed for multicast communication. The communication is inspired by the biological brain's neurons. The packets represent the neuron spikes that contain the information to be delivered from the source to the destination processing elements (Plana et al. 2008).

3.18 AUTONOMIC NETWORK ON CHIP USING THE BIOLOGICAL IMMUNE SYSTEM

The dynamic nature of bio-inspired solutions are being used in the NoC. To make the NoC autonomic, the self-configuration, self-healing and self-optimization characteristics of the immune system are used in the NoC. Application behavior and system state change (pathogens and infections) are detected by the bio-inspired NoC (BNoC) just like the immune system; the BNoC delivers a response to heal (adapts to change) it. In BNoC, self-configuration, self-healing and self-optimization are performed in the application, communication and architecture layers (Bakhouya 2010).

3.19 FAULT-TOLERANT NoC USING BIOLOGICAL BRAIN TECHNIQUES

The fault-tolerant characteristics of synaptogenesis and sprouting are biological brain techniques being implemented in the NoC to make communication reliable (Sethi, Hussin, and Hamid 2013b, 2013a, 2014, 2016). Synaptogenesis and sprouting are self-adapting and self-healing mechanisms, respectively, of the biological brain.

3.20 BIO-INSPIRED ONLINE FAULT DETECTION IN THE NoC INTERCONNECT

A bio-inspired fault-tolerant algorithm inspired by the working mechanism of biological neurons and synapses is used to detect faulty interconnects. The online bio-inspired algorithm works during run time as it detects the spatial and temporal faults during the NoC communication (McElholm et al. 2014).

3.21 BIO-INSPIRED SELF-AWARE NoC FAULT-TOLERANT ROUTING ALGORITHM

PSO is used in the bio-inspired routing algorithm to efficiently load-balance the traffic in the NoC in the presence of faults. The bio-inspired algorithm uses

Bio-Inspired Algorithms and Implementation 53

synchronous, asynchronous and self-aware capabilities of the PSO algorithm that enable the bio-inspired algorithm to make intelligent and dynamic decisions so as to have high efficiency in terms of throughput, power consumption and latency (Abba and Lee 2017).

REFERENCES

Abba, Sani, and Jeong-A Lee. 2017. "Bio-inspired self-aware fault-tolerant routing protocol for network-on-chip architectures using Particle Swarm Optimization." *Microprocessors and Microsystems* 51:18–38.

Bakhouya, Mohamed. 2010. "Towards a bio-inspired architecture for autonomic network-on-chip." *2010 International Conference on High Performance Computing & Simulation.*

Biyanto, Totok R, Sonny Irawan, Henokh Y Febrianto, Naindar Afdanny, Ahmad H Rahman, Kevin S Gunawan, Januar AD Pratama, and Titania N Bethiana. 2017. "Killer whale algorithm: An algorithm inspired by the life of killer whale." *Procedia Computer Science* 124:151–157.

Castro, Leandro Nunes, Leandro Nunes De Castro, and Jonathan Timmis. 2002. *Artificial immune systems: A new computational intelligence approach*: Springer Science & Business Media.

Chu, Shu-Chuan, Pei-Wei Tsai, and Jeng-Shyang Pan. 2006. "Cat swarm optimization." *Pacific Rim International Conference on Artificial Intelligence.*

Drigo, M. 1996. "The ant system: Optimization by a colony of cooperating agents." *IEEE Transactions on Systems, Man, and Cybernetics-Part B* 26 (1):1–13.

Eesa, Adel Sabry, Adnan Mohsin Abdulazeez Brifcani, and Zeynep Orman. 2013. "Cuttlefish algorithm-a novel bio-inspired optimization algorithm." *International Journal of Scientific & Engineering Research* 4 (9):1978–1986.

Heidari, Ali Asghar, Seyedali Mirjalili, Hossam Faris, Ibrahim Aljarah, Majdi Mafarja, and Huiling Chen. 2019. "Harris hawks optimization: Algorithm and applications." *Future Generation Computer Systems* 97:849–872.

Karaboga, Dervis, and Bahriye Akay. 2009. "A comparative study of artificial bee colony algorithm." *Applied Mathematics and Computation* 214 (1):108–132.

Liu, Lizheng, Yi Jin, Yi Liu, Ning Ma, Zhuo Zou, and Lirong Zheng. 2017. "Designing bio-inspired autonomous error-tolerant massively parallel computing architectures." *2017 30th IEEE International System-on-Chip Conference (SOCC).*

McElholm, Malachy, Jim Harkin, Junxiu Liu, and Liam McDaid. 2018. "Scalable Bio-inspired Fault Detection, Isolation and Recovery in NoCs." *2018 IEEE Symposium Series on Computational Intelligence (SSCI).*

McElholm, Malachy, Jim Harkin, Liam McDaid, and Snaider Carrillo. 2014. "Bio-inspired online fault detection in NoC interconnect," in *Energy-efficient fault-tolerant systems*, 241–267: Springer.

Mishra, Prabhakar, A Nidhi, and JK Kishore. 2014. "Novel bio-inspired cobweb topology for highly scalable and cost efficient networks on chip." *2014 IEEE International Conference on Electronics, Computing and Communication Technologies (CONECCT).*

Parpinelli, Rafael S, and Heitor S Lopes. 2011. "New inspirations in swarm intelligence: A survey." *International Journal of Bio-Inspired Computation* 3 (1):1–16.

Plana, Luis A, John Bainbridge, Steve Furber, Sean Salisbury, Yebin Shi, and Jian Wu. 2008. "An on-chip and inter-chip communications network for the spinnaker massively-parallel neural net simulator." *Second ACM/IEEE International Symposium on Networks-on-Chip (NOCS 2008).*

Sethi, Muhammad Athar Javed, Fawnizu Azmadi Hussin, and Nor Hisham Hamid. 2013a. "Implementation of biological sprouting algorithm for NoC fault tolerance." *2013 IEEE International Conference on Circuits and Systems (ICCAS).*

Sethi, Muhammad Athar Javed, Fawnizu Azmadi Hussin, and Nor Hisham Hamid. 2013b. "Synaptogenesis based bio-inspired NoC fault tolerant interconnects." *2013 IEEE International Conference on Control System, Computing and Engineering.*

Sethi, Muhammad Athar Javed, Fawnizu Azmadi Hussin, and Nor Hisham Hamid. 2014. "Bio-inspired NoC fault tolerant techniques." *2014 5th International Conference on Intelligent and Advanced Systems (ICIAS).*

Sethi, Muhammad Athar Javed, Fawnizu Azmadi Hussin, and Nor Hisham Hamid. 2016. "Biologically inspired network on chip fault tolerant algorithm using time division multiplexing." *2016 6th International Conference on Intelligent and Advanced Systems (ICIAS).*

Vogels, Werner, Robbert Van Renesse, and Ken Birman. 2003. "The power of epidemics: Robust communication for large-scale distributed systems." *ACM SIGCOMM Computer Communication Review* 33 (1):131–135.

Yang, Xin-She. 2010a. *Nature-inspired metaheuristic algorithms*: Luniver Press.

Yang, Xin-She. 2010b. "A new metaheuristic bat-inspired algorithm." In *Nature inspired cooperative strategies for optimization (NICSO 2010)*, 65–74: Springer.

Yang, Xin-She. 2012. "Flower pollination algorithm for global optimization." *International Conference on Unconventional Computing and Natural Computation.*

Yang, Xin-She, and Suash Deb. 2009. "Cuckoo search via Lévy flights." *2009 World Congress on Nature & Biologically Inspired Computing (NaBIC).*

4 Bio-Inspired NoC Fault-Tolerant Algorithms

4.1 BIOLOGICAL BRAIN CHARACTERISTICS

The biological brain is a complex organ with a network of neurons interconnected by synapses. It works faster than any supercomputer. The neuron is the basic building unit of the brain. There are around 80 to 120 billion neurons in the human brain (Hashmi et al. 2011). Figure 4.1 shows the biological neuron.

Neurons are connected through synaptic junctions that are formed by connections between the dendrites of one neuron and the axon terminal of another neuron, as shown in Figure 4.2. The input to neurons comes from dendrites, and the output follows through axons to the axon terminal. The axon is covered by a myelin sheet, which protects the signals traveling inside the axon. This sheet also increases the speed of the electrical signals traveling through the axon. The neuron's cell body is called the soma, which contains the nucleus. The neuron on one side of the junction is the presynaptic junction neuron, while the one on the other side is called a postsynaptic junction neuron. The synaptic cleft separates the axon terminal and dendrites. Neurons only receive a signal from those neurons through which they are connected. After summing all the inputs coming from various neurons, the neurons trigger the output if the signal strength is above a certain threshold. This output is in a chemical form as the electrical signal flows through the axon (Breedlove, Watson, and Rosenzweig 2007).

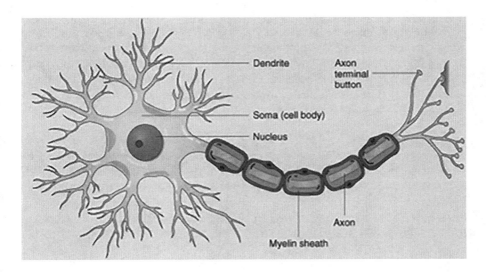

FIGURE 4.1 Biological neuron.

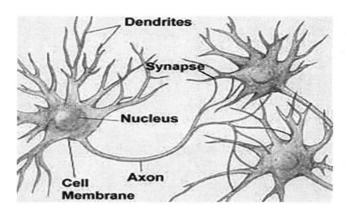

FIGURE 4.2 Biological synapse.

The brain is highly adaptable, and new connections (synapses) are dynamic in nature. There are between 1000 and 10,000 new connections per neuron in the brain. These connections can change very rapidly. Plasticity (changes in the brain) reduces with age. Damaged neurons self-destruct (a process called apoptosis) and are reabsorbed by the body. There are around five support cells for every neuron, called neuralgia or just glia, and these cells are responsible for maintenance issues. Some damaged neurons cannot be repaired in humans—for example, the severing of the spinal cord. If the firing rate of a neuron is higher than the resting rate, some neurons get fatigued and stop contributing to information flow. Then, due to the death of a neuron, other neurons take over and communication between them remains the same, but with a new path (Kalat 2015).

Two biological brain algorithms, namely synaptogenesis and sprouting, have been adopted and implemented to make Network on Chips (NoCs) fault-tolerant. These are self-adapting and self-healing mechanisms. These algorithms help the brain to recover when neurons and synapses get damaged. These bio-inspired NoC fault-tolerant algorithms are implemented using two connection setups or communications services. The two communication services are guaranteed throughput (GT) and best-effort services (BE).

4.2 SYNAPTOGENESIS

Synaptogenesis is a self-adapting mechanism in the human brain where two neurons attempt to connect and communicate with each other. In this phenomenon, the growth cone (having lamellipodium and filopodia) at the top of the axon and dendrites terminals, finds a path to a target neuron. The filopodia actually finds the path for a connection with the target neuron. A chemical attractant is released by the target neuron to attract the growth cone. A synapse is formed between the source and target neuron using this method (Breedlove, Watson, and Rosenzweig 2007). The synaptogenesis process is shown in Figure 4.3.

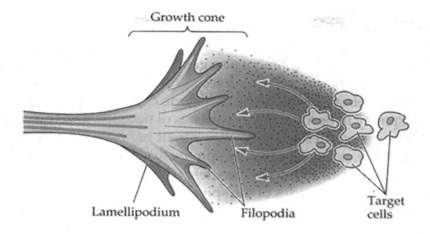

FIGURE 4.3 Synaptogenesis.

4.3 SPROUTING

Sprouting is the self-healing mechanism of the brain. Due to various reasons, including blood clots, hypertension and high blood pressure, neurons and synapses sometimes get damaged. In these situations, a new synapse or sprout emerges from the axon, which is already connected to the source neuron. With the help of chemical adhesion material (CAM), this sprout is connected with the target neuron, as shown in Figure 4.4. After various experiments on fishes and frogs (amphibians), it is concluded that this feature is not present in humans (Breedlove, Watson, and Rosenzweig 2007).

4.4 BIO-INSPIRED NoC ALGORITHMS

The biological brain's algorithms were adopted in the NoC to make it fault-tolerant. In the first phase, a synaptogenesis algorithm was implemented in the NoC, and later the algorithm was improved by adopting the sprouting concept. In the coming sections, the improved algorithm is called the "sprouting-based" algorithm.

4.4.1 Synaptogenesis-Based NoC

This algorithm helps to establish an optimal connection between the source processing element (PE) and destination PE. It is able to detect the static faults during the establishment of a connection. During the communication between source and destination, if any interconnect becomes faulty, a new synapse is formed from the neighboring router. This is possible because the destination PE and source PE are constantly communicating with each other. The destination address in the flits helps to recover from the faults. In synaptogenesis, a credit mechanism is used to detect

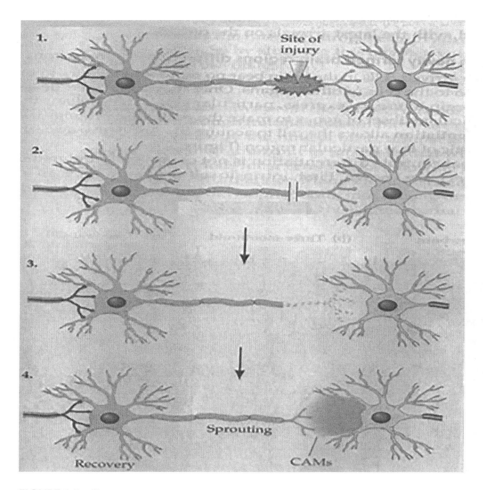

FIGURE 4.4 Sprout.

the fault in the NoC. It takes 3 ns to detect the fault, as the router does not receive the credit packet from the neighboring router. In order to improve fault detection and performance of the NoC, a sprouting algorithm was proposed.

4.4.2 SPROUTING-BASED NOC

With the help of the sprouting algorithm, a synapse (sprout) emerged from the synapse already connected with the source PE when an interconnect became faulty during the communication. This algorithm detects the run-time faults during the communication and creates a workaround synapse (connection) to bypass the faulty interconnects. The sprouting algorithm detects the faults as the flits arrive at the block port of a particular router. The port is blocked when a fault occurs in the

Bio-Inspired NoC Fault-Tolerant Algorithms

interconnect. No credit packet is required in the sprouting algorithm to detect the fault, avoiding the waste of 3 ns, as in the case of synaptogenesis. This helps the algorithm to quickly create a workaround synapse compared to synaptogenesis. The sprout emerges as the destination address is saved in every flit. Therefore, the overall performance of the sprouting algorithm is better compared to the synaptogenesis algorithm.

The algorithms recover from the static and run-time faulty interconnects by creating a workaround synapse. Multiple synapses are formed between the source PE and destination PE, which efficiently utilizes the bandwidth, maximizes the throughput and recovers from the faults quickly.

4.5 BIO-INSPIRED NoC FRAMEWORK

Initially, the synapse is initiated from the source PE toward the destination PE. The synapse is routed on the NoC using two-hop information and arrives at the destination. The destination in return sends the route reply packet. After receiving the route reply packet at the source PE, the flits are sent from the source PE to the destination PE. If no fault is detected during the communication, then flits reach the destination. Otherwise, the neighbor router detects the fault and initiates the synapse connection (sprout) between the current router and either the older synapse or the destination router. If there is no older synapse to connect with, the sprout is directly connected with the destination PE.

Whenever a fault occurs, the destination address, routing path and routers traversed are saved in the new synapse. These parameters help the newer synapse connect with the older synapse. The bio-inspired algorithm is robust, as it tries to connect with the older synapse and also avoids the traversal of unnecessary routers. This also decreases the latency of the packets/flits even when faults occur. The bandwidth of the NoC is also efficiently utilized, and throughput is slightly decreased for a short period of time as the NoC recovers from the faulty interconnect. During the network recovery time, the destination is still receiving the flits from other synapses. These flits have already traversed the faulty interconnect. The generalized framework of the bio-inspired algorithm is shown in Figure 4.5.

In bio-inspired algorithms, the "Scheduler" module of the port plays a significant role in fault detection. The port is a fundamental building block of the router. The routers are connected with each other through interconnects in an NoC network.

4.6 BIO-INSPIRED NoC NETWORK

In the bio-inspired NoC, routers are connected with each other through bidirectional links. Each PE is connected with the router through a network interface. There are five ports in the router. Four ports are connected with the neighbor routers, that is, north, east, west and south, while the fifth port is connected with the PE. Every port has an "inPort", which receives packets from the neighbor router or from the PE. The "OutPortCalc" is used to calculate the optimum output port for "synap" packets. The output port is determined based on the destination and faulty situation of neighbor router, collected with the help of control packets. The "Scheduler" is used to control

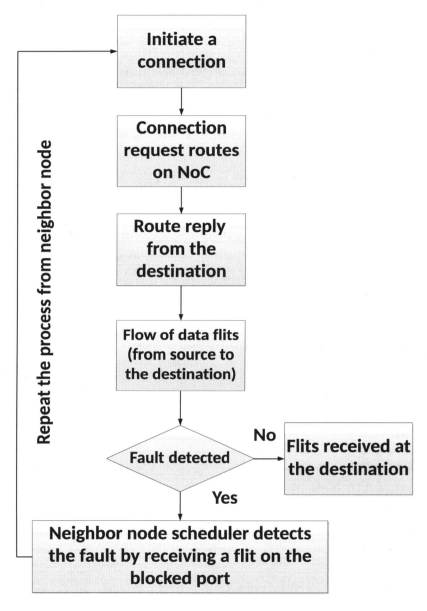

FIGURE 4.5 Bio-inspired NoC algorithm process flow.

the output port (Ben-Itzhak et al. 2012), as shown in Figure 4.6. The "Scheduler" sends the packet at the (Out) link when the interconnect is free. The "Scheduler" is connected with the "inPort" for disseminating the fault detection information and data flits to the "inPort" of that particular port. Later, these data flits are sent on the new workaround synapse initiated by "inPort".

Bio-Inspired NoC Fault-Tolerant Algorithms

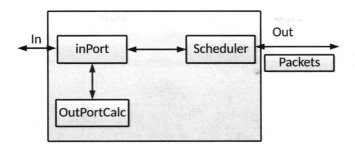

FIGURE 4.6 Port architecture.

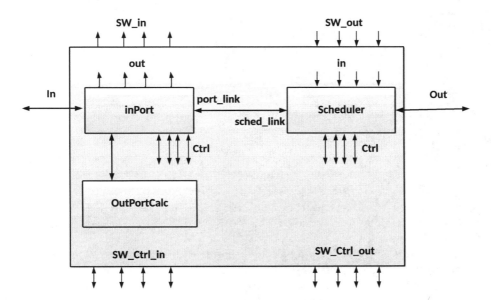

FIGURE 4.7 Detailed port architecture.

The detailed architecture of the port is shown in Figure 4.7. The "SW_in", "SW_out", "SW_Ctrl_in", "SW_Ctrl_out" and "Ctrl" pins are used to control the flow of data flits between routers. "SW_in" and "SW_out" are unidirectional lines, whereas "SW_Ctrl_in" and "SW_Ctrl_out" are bidirectional lines. Control packets of grant, request, response and nack are shared through the "SW_Ctrl_in" and "SW_Ctrl_out" pins between the ports of a particular router. Similarly, "SW_in" and "SW_out" pins are used to send data flits between various ports of the router.

The "out" pins of the "inPort" are connected to the "SW_in" pins of the port, while "Ctrl" pins are connected with the "SW_Ctrl_in" pins. The "in" pins of the "Scheduler" are connected with "SW_out" pins, while "Ctrl" pins are connected with "SW_Ctrl_out" pins, as shown in Figure 4.8. Each "SW_in" pin is connected with every port in the router. Similarly, "SW_out" pins of the particular port are connected with every other port in the router. The "SW_Ctrl_in" pins in the router

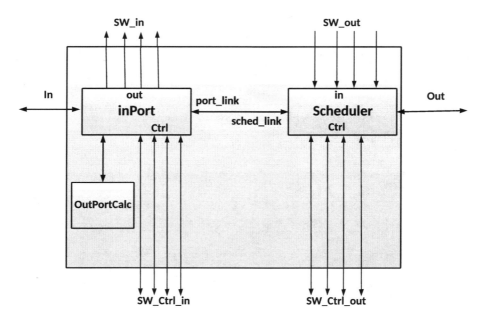

FIGURE 4.8 Detailed port architecture connections.

are connected with every other port in the router through "SW_Ctrl_out". Similarly, "SW_Ctrl_out" pins are connected with "SW_Ctrl_in" pins of every other port on the particular router.

Each of the five ports in a router are connected with each other through crossbar interconnections. Every port is connected with every other port in the router, as shown in Figure 4.9 and Figure 4.10. The "SW_in", "SW_out", "SW_Ctrl_in" and "SW_Ctrl_out" are connected together between different ports of the router. In Figure 4.9, four pins of "SW_out" of port 0 (north) are connected with the one "SW_in" pin of port 1 (west), port 2 (south), port 3 (east) and port 4 (local core), respectively. Similarly, four pins of "SW_Ctrl_out" of port 0 are connected with the each "SW_Ctrl_in" of ports 1, 2, 3, and 4 respectively, as shown in Figure 4.10. The same mechanism is used for other ports of a router. The data flits traverse from "SW_in" of one router port to "SW_out" of the other port of the same router, whereas control packets traverse on the "SW_Ctrl_in" and "SW_Ctrl_out" (Ben-Itzhak et al. 2012).

4.7 BIO-INSPIRED NoC FAULT-TOLERANT ALGORITHM

In the NoC, routers discover themselves at system initialization. The network discovery is initiated by the "Scheduler" of the router port. The Scheduler sends the request packet to the neighbor router, which in return, replies with a response control packet. The response packet is sent by the "inPort" module of the particular router port. After sharing the control packets, routers know about the status of their neighbor routers, that is, either they are working or not working. If after 1 ns the response is

Bio-Inspired NoC Fault-Tolerant Algorithms

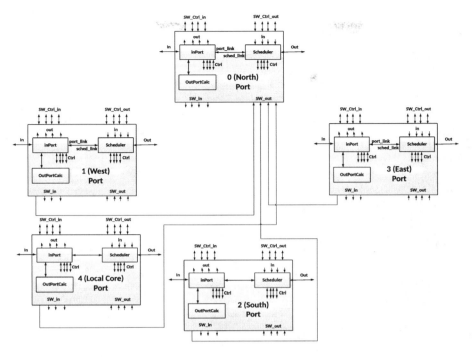

FIGURE 4.9 Router port connections for data paths (SW_in with SW_out).

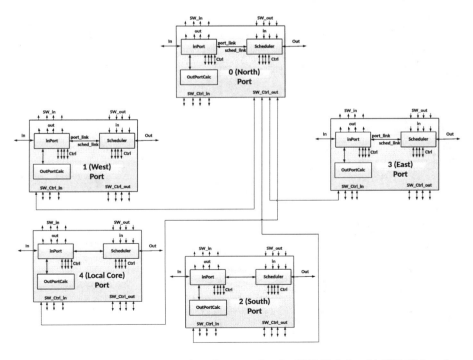

FIGURE 4.10 Router port connections for control paths (SW_Ctrl_in with SW_Ctrl_out).

not received at the "Scheduler" port of the sending router, then that port is blocked, as either the interconnect or neighbor router is faulty. The bio-inspired NoC algorithm only needs one network discovery at initialization. This efficiently utilizes the bandwidth of the NoC and improves overall performance.

After network discovery, the source PE initiates the synapse "synap" packet to connect with the destination PE. When a router receives the "synap" packet, it sends information set (IS) request packets to its neighbors through the "Scheduler". Neighbor routers reply to this request with IS response packets containing its information (direction neighbor, DN) and its immediate neighbors (IN) information through the "inPort". The immediate neighbors for an ISW (west) packet are east, north and south, while west, north and south are the INs of ISE (east) packets. Similarly, north, east and west are the immediate neighbors of ISS (south) packets, while ISNs (north) have the INs of the south, east and west. The neighbor router sends back the information in the ISW, ISE, ISS and ISN response packets, as shown in Figure 4.11. The information in the IS response packet is 1 (working), 0 (not working) and −1 (neighbor does not exist). The flow of the IS request and response packets

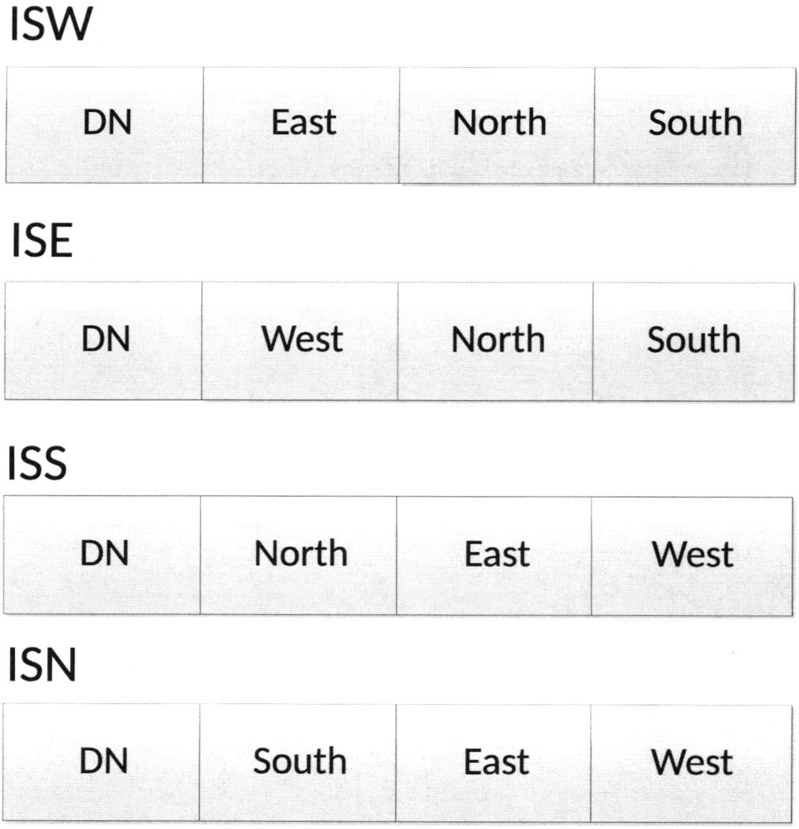

ISW

DN	East	North	South

ISE

DN	West	North	South

ISS

DN	North	East	West

ISN

DN	South	East	West

FIGURE 4.11 IS response packet format.

Bio-Inspired NoC Fault-Tolerant Algorithms

is shown in Figure 4.12 and Figure 4.13 with router id=10 as an example (Nicopoulos, Narayanan, and Das 2009).

Upon receiving all the IS packets from the neighbors, the router prioritizes the path based on the information present in the IS packets and the location of the destination, as shown in Figure 4.14. For example, if the destination (PE connected with router 16) is toward the east, depending on the faulty situation of the neighbors, the first priority to send the packet is toward the east, the second priority is toward the south, the third priority is toward the north and the last option is to send the packet toward the west.

If DN is zero (not working), then that path is blocked immediately. If the DN is 1 (working) and all three INs are faulty, the path is still blocked, as the "synap" will not be able to pass through the neighbor routers. If the DN is 1 and any one of the INs is working (1), the "synap" packet is sent in that particular direction. This algorithm can be further optimized by controlling the flow of the packet; the "synap" packets should be sent in specific directions where more INs are working, rather than in one direction only.

The following are the five broad cases that may occur based on the information received and processed by the router (outPortCalc) using the information present in ISW, ISE, ISS and ISN packets and based on the location of the destination.

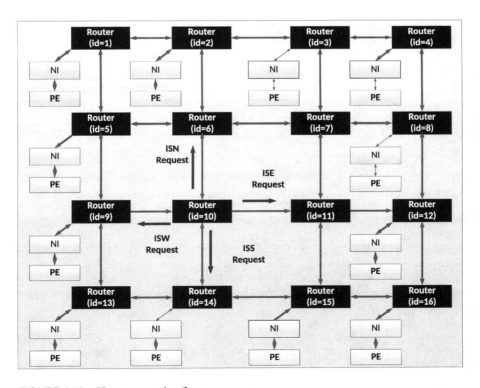

FIGURE 4.12 IS request packet flow.

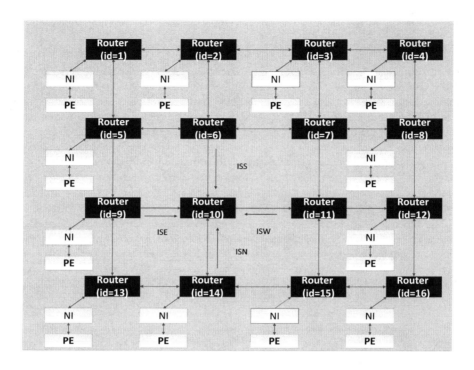

FIGURE 4.13 IS response packets.

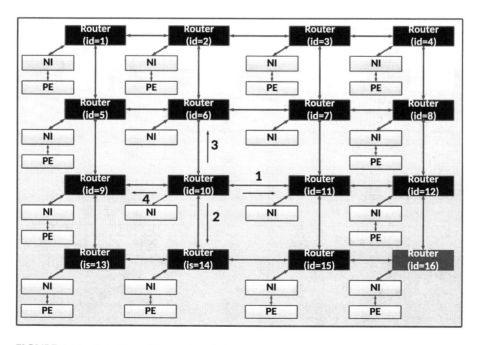

FIGURE 4.14 Priorities of "synap" packet.

Bio-Inspired NoC Fault-Tolerant Algorithms

CASE 1: DESTINATION IS TOWARD THE EAST

If the destination is toward the east, then the first priority is to send the "synap" packet in this direction. If the east interconnect is faulty, then the next priority will be to send the flit toward the north. If the north interconnect is faulty, then the next priority is to send the "synap" packet toward the south. The last priority will be to send the "synap" packet toward the west, which is opposite from the destination. If all ports are faulty, then the "synap" packet has to wait.

CASE 2: DESTINATION IS TOWARD THE WEST

If the destination is toward the west, then the first priority is to send the "synap" packet in this direction. If the west interconnect is faulty, then the next priority will be to send the flit toward the north. If the north interconnect is faulty, then the next priority is to send the "synap" packet toward the south. The last priority will be to send the "synap" packet toward the east, which is opposite from the destination. If all ports are faulty, then the "synap" packet has to wait.

CASE 3: DESTINATION IS TOWARD THE NORTH

If the destination is toward the north, then the first priority is to send the "synap" packet in that direction. If the north interconnect is faulty, then the next priority will be to send the flit toward the east. If the east interconnect is faulty, then the next priority is to send the "synap" packet toward the west. The last priority will be to send the "synap" packet toward the south, which is opposite from the destination. If all ports are faulty, then the "synap" packet has to wait.

CASE 4: DESTINATION IS TOWARD THE SOUTH

If the destination is toward the south, then the first priority is to send the "synap" packet in that direction. If the south interconnect is faulty, then the next priority will be to send the flit toward the east. If the east interconnect is faulty, then the next priority is to send the "synap" packet toward the west. The last priority will be to send the "synap" packet toward the north, which is opposite from the destination. If all ports are faulty, then the "synap" packet has to wait.

CASE 5: LOCAL CORE (PE)

If the current router's x-axis (rx) and y-axis (ry) are equal to the target router's (destination PE) x-axis (dx) and y-axis (dy), then the "synap" packet is send toward the local core (PE).

Figure 4.15 shows the x-axis and y-axis values of routers in a 4×4 NoC. The router (0) has rx=0 and ry=0 values, whereas router (15) has rx=3 and ry=3 values. The dx and dy values also depend on the coordinate values of the routers. For example, if the destination is at the router (10), then the value of dx and dy will be 2. The details regarding different possibilities of the routing algorithm when the destination is toward the east, west, south and north are shown in Figure 4.16 to Figure 4.19,

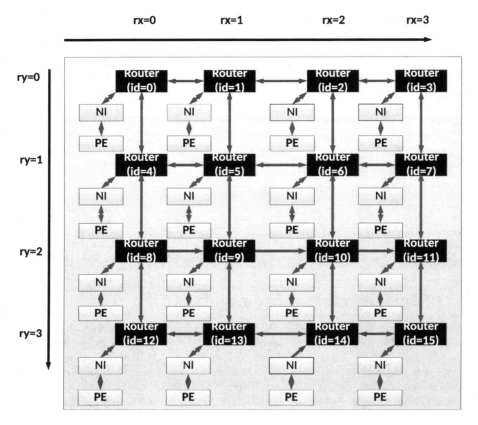

FIGURE 4.15 The rx and ry values of routers in 4 × 4 NoC.

respectively. In these figures, dy, dx specify the destination address in the form of coordinates, while rx, ry specifies the current router location in the NoC using x-axis and y-axis values.

It can be seen in Figure 4.16 that the routing algorithm depends on the values of dx, dy, rx and ry. If dx>rx, then the destination is toward east. The first priority is to send the "synap" packet in this direction, but if the east port is blocked due to a faulty interconnect, then the second priority depends on the location of the destination. If the destination is toward the south or at the same column of the NoC (dy>ry || dy==ry), then the "synap" packet is traversed toward the south first and later toward the north if the route toward the south is faulty. Or if the destination is toward the north (dy<ry), then the first priority will be to send the packet toward the north and later to the south if the north path is faulty. The last priority will be to send the packet toward the west, which is opposite from the destination. If all ports are faulty, then the "synap" packet has to wait.

Figure 4.17 shows that if dx<rx, then the destination is toward the west. The first priority is to send the "synap" packet in this direction, but if the west port is blocked due to a faulty interconnect, then the second priority depends on the location of the

Bio-Inspired NoC Fault-Tolerant Algorithms

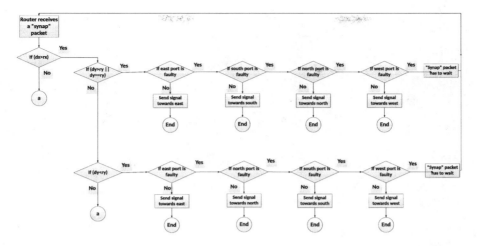

FIGURE 4.16 Process flow when the destination is toward the east, with various possibilities.

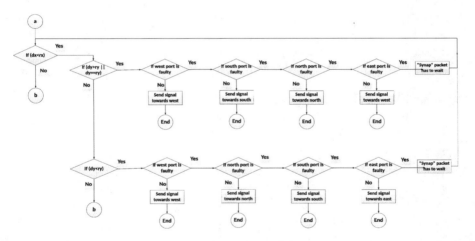

FIGURE 4.17 Process flow when the destination is toward the west, with various possibilities.

destination. If the destination is toward the south or at the same column (dy>ry || dy==ry), then the "synap" packet is traversed toward the south first and later toward the north if the route toward the south is faulty. Or if the destination is toward the north (dy<ry), then the first priority will be to send the packet toward the north and later to south if the north path is faulty. The last priority will be to send the packet toward the east, which is opposite from the destination. If all ports are faulty, then the "synap" packet has to wait.

As shown in Figure 4.18, if dy>ry, then the destination is toward the south. The first priority is to send the "synap" packet in this direction, but if the south port is

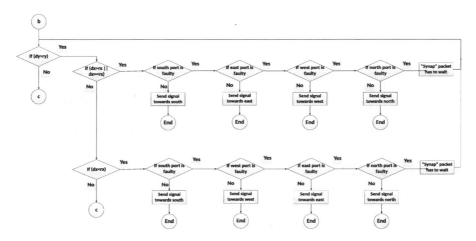

FIGURE 4.18 Process flow when the destination is toward the south, with various possibilities.

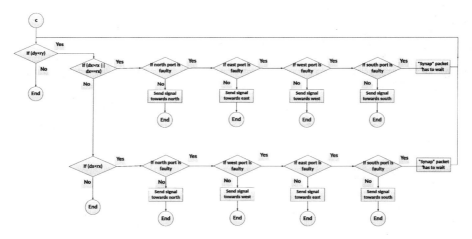

FIGURE 4.19 Process flow when the destination is toward the north, with various possibilities.

blocked due to a faulty interconnect, then the second priority depends on the location of the destination. If the destination is toward the east or at the same row (dx>rx || dx==rx), then the "synap" packet is traversed toward the east first and later toward the west if the route toward the east is faulty. Or if the destination is toward the west (dx<rx), then the first priority will be to send the packet toward the west and later to the east if the west path is faulty. The last priority will be to send the packet toward the north, which is opposite from the destination. If all ports are faulty, then the "synap" packet has to wait.

As shown in Figure 4.19, if dy<ry, then the destination is toward the north. The first priority is to send the "synap" packet in this direction, but if the north port is

Bio-Inspired NoC Fault-Tolerant Algorithms

blocked due to a faulty interconnect, then the second priority depends on the location of the destination. If the destination is toward the east or at the same row ($dx>rx$ || $dx==rx$), then the "synap" packet is traversed toward the east first and later toward the west if the east path is faulty. Or if the destination is toward the west ($dx<rx$), then the first priority will be to send the packet toward the west and later to the east if the west path is faulty. The last priority will be to send the packet toward the south, which is opposite from the destination. If all ports are faulty, then the "synap" packet has to wait.

After the establishment of a synapse between the source PE and destination PE, data flits are sent from the source to the destination over the path specified by the "synap" packet. The path is saved in every flit to address fault tolerance. If there is no faulty interconnect along the way, then the data flits will traverse the path. If during the communication an interconnect becomes faulty, a neighbor router detects the fault through the "Scheduler". A new sprout (synapse) emerges from the neighbor router "inPort" to bypass the faulty interconnect. The port of the router connected with that particular faulty interconnect is blocked through the "Scheduler" and the "sourceInform" packet is sent to "inPort". The "sourceInform" is used to inform and initiate a new workaround synapse from the current router to bypass the faulty interconnect or router, as shown in Figure 4.20.

```
    cModule* curRouter = getParentModule()->getParentModule();
  cModule* curPort = getParentModule();
    int portN=curPort->getIndex();

//EV<<"Informing inPort to initiate a new work-around
synapse"<<endl;
    sourceInform=new NoCInformSourceMsg("Inform Source");
    sourceInform->setKind(NOC_INFORMSOURCE_MSG);
    sourceInform->setSrcId(srcId);
    sourceInform->setDstId(dstId);
    sourceInform->setSynpseId(synpseID);
    sourceInform->setTotalCount(totalCount);

        for (int i=0; i<totalCount;i++)
        {
            int outPort=msg->getSynpOutPort(i);
            sourceInform->setSynpOutPort(i,outPort);
        }

        for (int i=0; i<totalCount;i++)
        {
            int routerIndex=msg->getRouterIndex(i);
            sourceInform->setRouterIndex(i,routerIn
        }

        //EV<<"Sending a packet to inPort from
  Scheduler"<<endl;
        send(sourceInform,"sched_link$o");
```

FIGURE 4.20   Source inform code.

The difference between "synaptogenesis" and "sprouting" algorithms is detecting and informing the "inPort" about faults. In synaptogenesis, it takes 3 ns to detect faults and to send "sourceInform" packets to "inPort" and initiate a new synapse packet (steps 16 and 17 of algorithm 1). This concept was improved with sprouting by changing the logic of the fault detection code. The "sourceInform" packet is sent to "inPort" as soon as the detection of the fault occurs (steps 9 and 10) of algorithm 2. The pseudo-code (algorithm 1 and algorithm 2) and the actual code in Appendix B of the "Scheduler" specify the exact differences between these bio-inspired algorithms. The actual code is shown for only the north port (portNumber==0). The same code is repeated for the other four ports with changes of port numbers and port names.

In the synaptogenesis algorithm, faulty interconnects are detected after the flit is sent to the neighbor router and when credit is not received from the neighbor router in 3 ns. The timer is used to waiting for 3 ns for the credit response from the neighbor router. If the credit is not received, then it means either the neighbor router or the interconnect is faulty (steps 6 to 18 of algorithm 1). In the sprouting algorithm, this timer is not used for optimization and the implementation of the sprout concept. Rather than detecting a faulty interconnect after sending the flit and waiting for credit response, the sprouting algorithm checks for the faulty interconnect before sending the flit to the out link (steps 6 to 15 of algorithm 2). This helps the sprouting algorithm avoid wasting 3 ns for the timer.

For example, in the synaptogenesis algorithm, if the flit arrives at the "Scheduler" at 8 ns and the port is not blocked due to a fault, then the credit packet from the neighbor will arrive at 10 ns. Even if the fault occurs at 9 ns, the neighbor router will detect the fault at 11 ns, as it will not receive the credit packet (after 2 ns, the credit packet will be sent by the neighbor router, while 1 ns is added so that it is received at the neighbor router successfully). In the case of the sprouting algorithm, as the fault detection does not depend on the timer and credit packet, it will be detected at 9 ns (at the arrival of flit) at the neighbor router. The benefit of the sprouting algorithm is that the faulty interconnect is detected as soon as possible when the fault occurs and when the flit is received at the "Scheduler".

Another benefit of the sprouting algorithm compared to the synaptogenesis algorithm is that it does not need a buffer to store the flits in a separate queue. The buffer is required in synaptogenesis to store the flits, as the fault is detected after sending the flits and when the router does not receive the credit packet after 3 ns. In that case, the "Scheduler" should have the copy of the flit so that the flit can be sent on the newer workaround synapse.

Algorithm 1: Synaptogenesis algorithm (pseudo-code for fault detection)

```
if (router port is already blocked)
{
Send only flit to inPort so that it can be sent on newer workaround
synapse;
}
else
```

Bio-Inspired NoC Fault-Tolerant Algorithms

```
{
Send flit on (Out) link of "Scheduler" to neighbor router;
Wait for 3 ns for credit packet; \\timer
if (credit received)
{
Cancel timer;
Increment credit at "Scheduler";
}
else
{
Fault detected;
Inform inPort to initiate newer workaround synapse to bypass
fault; \\sourceInform packet
}
}
```

Algorithm 2: Sprouting algorithm (pseudo-code for fault detection)

```
if (router port is already blocked)
{
Send only flit to inPort so that it can be sent on newer
workaround synapse;
}
else
{
if (this port is blocked)
{
Fault detected;
Inform inPort to initiate newer workaround synapse to bypass
fault; \\sourceInform packet
}
else
{
Send flit on (Out) link of "Scheduler" to neighbor router;
}
}
```

Algorithm 3 is the pseudo-code of the overall bio-inspired algorithm. This algorithm only specifies one case when the destination is toward the east. Initially, a synapse connection is established between the source PE and destination PE (steps 1 to 14). The "synap" packet is initiated from the source PE, and it traverses through the routers by having two-hop neighbors' information (steps 1 to 13). Once the synapse (connection) is established between the source PE and destination PE, flits traverse over this synapse (steps 14 and 15). If during the communication, the interconnect becomes faulty, then the neighbor router detects the fault and initiates the work-around synapse (sprout) connection (steps 16 to 18). During establishment of the new workaround synapse, the "synap" packet always tries to connect with the older synapse. If an older synapse connection does not exist, then the "synap" packet is connected directly with the destination PE. After the workaround synapse connection is established, data flits move over it (steps 19 and 20).

Algorithm 3: Bio-inspired algorithm (when the destination is toward the east)

```
"Synap" packet initiated from source to destination.
if ("Synap" reached at Destination)
Send the "Synap" packet to destination core
Else if (Destination toward East)
check East DN and INs condition
Else if //East DN or INs Faulty
check North DN and INs condition
Else if //North DN or INs Faulty
Check South DN and INs condition
Else if //South DN or INs Faulty
Check West DN and INs Condition
Else if //if all neighbor ports of router are faulty
"Synap" packet has to wait
Send the "Synap" packet to selected direction
Synapse connection is formed between source and destination.
Data flits move from source to destination over the synapse.
Upon interconnect failure neighbor router detects the fault.
The flits are saved in the "Scheduler" and later at the
"inPort" module of the router.
Router initiates a synap connection (sprout) and tries to
connect the new synapse with the older synapse.
If older synapse does not exist then the new synapse directly
connects with the destination.
Data flits are sent from the workaround newer synapse.
```

Figure 4.21 shows the implementation of the bio-inspired algorithm when interconnect failure occurs. The synapse was constructed from the source (0) to destination (15) at the start of the communication between them. During the simulation, the links between various routers became faulty and a workaround synapse was initiated to bypass the faulty interconnects. The temporary synapse was formed between router (1) and router (7) to bypass the faulty interconnect between router (1) and router (2). As shown in the figure, the workaround synapse is bypassing router 2 and router 3 and connecting directly with the old synapse at router 7. The same mechanism is adopted for any number of faults in the NoC. The crosses in the figure show the faulty interconnects.

4.8 BIO-INSPIRED BE AND GT NoC ALGORITHM AND ARCHITECTURES

The bio-inspired NoC algorithms are implemented using BE and GT and combined BE-GT services. In both service levels, a 4×4 NoC mesh topology is used. The routers are connected with each other, while the PEs are connected with routers through the network interface (NI). The overall port architecture of routers in both NoCs is

Bio-Inspired NoC Fault-Tolerant Algorithms

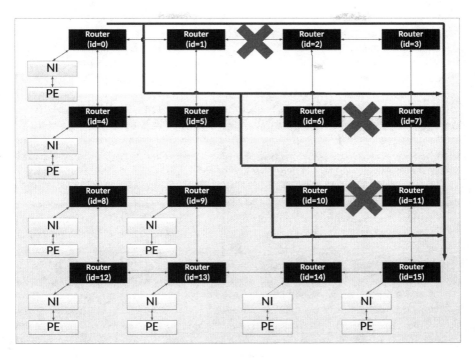

FIGURE 4.21 Multiple connections between source and destination.

the same except the architecture of the "Scheduler", "inPort" and some control/data pins. These control and data pins are used in the BE and GT architecture to control the flow of flits between routers. In the GT architecture, there are time slots at the "Scheduler" and multiple queues at the "inPort". However, in the BE architecture, there is a credit-based flow control technique to manage the communication between routers.

4.8.1 BE Services in the NoC Architecture

The port architecture of the BE router is similar to Figure 4.6. In BE services, no resources (routers, interconnects) are reserved. Any source can send the packet based on the availability of an output port. The decision to send the packet in a particular direction is made on a router-to-router basis based on the destination address. The "Scheduler" detects the faults, and later it transfer flits to the "inPort" to save these flits. The "inPort" contains the buffers to store these flits.

In BE services, credit-based flow control (Hansson, Goossens, and Rădulescu 2007) is used to control the flow of flits between routers and PEs. In this mechanism, the "Scheduler" looks at the credit counter before sending the flit to the output port. If the credit is more than zero, then it can send the flits at that particular port; a zero credit means the port is busy and waiting for feedback (credit) from a neighbor. The exchange of credit information protects the links from congestion and collision and avoids the deadlock situation.

After receiving a flit at the "inPort", the credit packet is sent back to the "Scheduler" to increment the virtual channel credit counter. The credit counter is incremented by one, as only one flit is sent by the Scheduler per clock cycle. Currently, two virtual channels are supported between routers to send flits on.

The code shown in Figure 4.22 is used to send the credit packet from "inPort" to the neighbor router "Scheduler". crd is the credit object created at the "inPort". The credit is incremented for a specific virtual channel (vc) by a number of flits (numFlits) at the "Scheduler". The "inPort" specifies the (vc) and (numFlits) parameters.

At the receiver side, this credit packet is received at the "Scheduler", and the credit counter is incremented by one, as specified by the number of flits ("numFlits") variable, as shown in Figure 4.23.

The flow of control and data packets is shown in Figure 4.24. The "synap" packet is initiated by the source PE to connect and start communication with the destination PE.

In BE services, before sending the flits to the next router, the particular port of a router sends a "request" packet to the port. The request packet specifies the target port that the particular port wants to send flits to. If the credit counter is greater than one and the port is not busy, then the target port will send the grant packet to the port. Then, the port will send the flits to the target port. If the port is busy or the credit counter is zero, then the target port will not send an acknowledge packet, and the port has to send the request again to the target port. Once the router sends a flit to the neighbor router, the credit counter is decremented, and the port status becomes busy until it receives the acknowledgment in the form of a credit packet from the neighbor router. If the router does not receive the credit by a specific time (3 ns), then

```
char credName[64];
sprintf(credName, "cred-%d-%d", vc,
numFlits);
    NoCCreditMsg *crd = new
NoCCreditMsg(credName);
    crd->setKind(NOC_CREDIT_MSG);
    crd->setVC(vc);
    crd->setFlits(numFlits);
    send(crd, "in$o");
```

FIGURE 4.22 Code for sending credit packet from inPort.

```
    int vc = msg->getVC();
    int num =
msg->getFlits();
    credits[vc] += num;
```

FIGURE 4.23 Code for incrementing credit counter at "Scheduler".

Bio-Inspired NoC Fault-Tolerant Algorithms

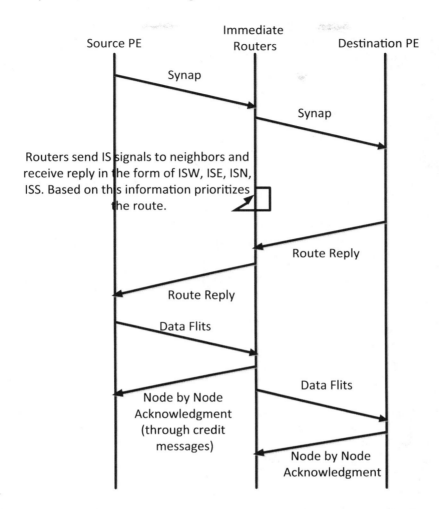

FIGURE 4.24 Timing diagram of packet flow between source PE and destination PE using BE.

the router will assume that a fault has occurred in the neighbor and will initiate a workaround synapse packet to bypass the fault.

In the BE routing algorithm of the NoC architecture, before making any decision regarding the direction of the "synap" packet, the information in the ISs is processed. The router makes sure that the direct neighbor is not faulty and that at least one IN is working. The code in Figure 4.25 explains the concept.

In this code, the destination is assumed to be toward the east. The same concept and logic are used for making decisions regarding the other directions. DNW is the variable that specifies the processed information of the ISW packet. The values are in the range of 0, 1, 2 and 3. Zero specifies that the DN and all INs are working.

```
if (DNW==0 &&
        isWest[srcId][1]==1 &&
        isWest[srcId][0]!=0 &&
        eastPort!=-1)
  {
        swOutPortIdx=eastPort;
  }
else if (DNW==0 &&
        isWest[srcId][2]==1 &&
        isWest[srcId][0]!=0 &&
        eastPort!=-1)
  {
        swOutPortIdx=eastPort;
  }
else if (DNW==0 &&
        isWest[srcId][3]==1 &&
        isWest[srcId][0]!=0 &&
        eastPort!=-1)
  {
        swOutPortIdx=eastPort;
  }
```

FIGURE 4.25 Code to process information in ISs.

DN values are calculated by summing up the "0" (not working status) and "–1" (neighbor does not exist) in IS packets. A 1 specifies that either the IN or DN is not working. Similarly, 2 and 3 specify the faulty interconnects in the neighbors of the current router. For DNW to equal 0, there are three different cases to check before making any decision. If the DN is 0 (faulty) (that is, the DN should not be equal to 0, *isWest*[srcId][0]!=0), then this path is not selected. For the other three cases, all the INs are checked (*isWest*[srcId][i]==1). If any one of the INs is working, then that path is selected. It is possible that the east port is already blocked due to a faulty interconnect, and it should be checked (eastPort!=-1).

Similarly, the same code is used for every NoC direction and for every value of DNW, DNS, DNE and DNN. The values of DNW, DNE, DNS and DNN range from 0 to 3. If the DNW, DNE, DNS and DNN value is 4, it means that the DN and all INs are not working (0) or the neighbor does not exist (–1). That NoC direction is totally avoided in that case. The code shown in Figure 4.26 is used to activate the particular interconnect during the simulation. The code is used to bring the particular interconnect back to a working condition by activating its ports.

The NoCRouterAliveMsg object is used to reactivate a particular port. The interconnects are connected with these ports. The curRouter object refers to the current router whose ports need activation. portNumber refers to the current port that needs activation, and curRouter->**par**("northPort_W")= 1, will activate the northPort back to working condition. The same logic is used for the other ports. Figure 4.27 shows

Bio-Inspired NoC Fault-Tolerant Algorithms

```
void SchedAsync::handleRouterAliveMsg(NoCRouterAliveMsg*
msg)
{

    cModule* curPort = getParentModule();
    portNumber = curPort->getIndex();

    if (portNumber == 0) {
        curRouter->par("northPort_W") = 1;
    } else if (portNumber == 1) {
        curRouter->par("westPort_W") = 1;
    } else if (portNumber == 2) {
        curRouter->par("southPort_W") = 1;
    } else if (portNumber == 3) {
        curRouter->par("eastPort_W") = 1;
    } else {
        curRouter->par("corePort_W") = 1;
        }
    }
}
```

FIGURE 4.26 Code to activate the ports.

the flow of packets when a router port becomes busy. During the transmission of packets, two sources want to communicate through one port; thus, one source has to wait to acquire the port. As can be seen in the figure, when source 2 sends a request packet to acquire the port, which is already acquired by source 1, source 2 will receive the "Port Busy" packet from the router through the "Ctrl" port and then it will wait for the "Port Free" packet before sending the packets through the "Ctrl" ports of the particular port of the router.

4.8.2 GT Services in the NoC Architecture

The port architecture of the GT is similar to Figure 4.6. The difference is between the architecture of the "inPort" and "Scheduler". Each "inPort" has a separate queue for every virtual channel. This avoids head of line (HOL) blocking whenever faults occur. The Scheduler has four time slots, which are allocated at run time to various virtual channels. The optimum four slots are reserved based on the literature (Hansson, Subburaman, and Goossens 2009). The time slots are managed and read by the time division multiplexing (TDM) clock. The TDM clock is running autonomously at every router, and the clock period is 2 ns. The time slots help to efficiently utilize and divide the bandwidth of the NoC among the four connections.

Different slots are allocated to four virtual channels: A, B, C and D. The routers (R) increment the slot number by 1, as stated in Equation (4.1). At every time slot, the router is sending one packet to the (Out) link from any virtual channel. This is

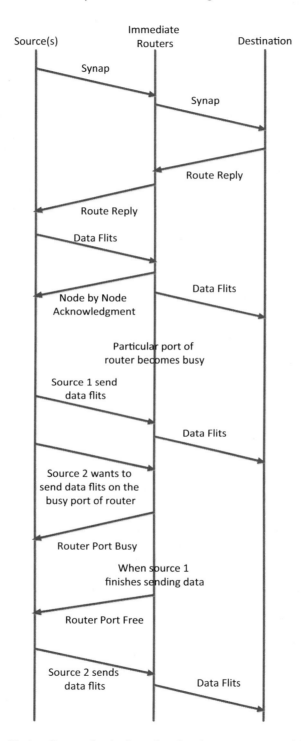

FIGURE 4.27 Timing diagram for the flow of packets between source and destination when router port is busy.

Bio-Inspired NoC Fault-Tolerant Algorithms

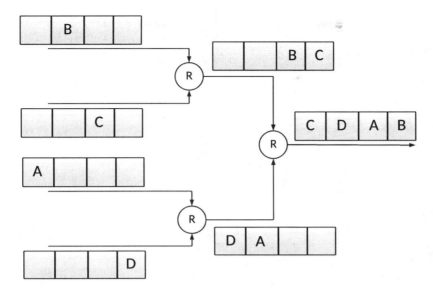

FIGURE 4.28 TDM mechanism for contention-free routing.

contention-free routing, as there is no collision between packets from different virtual channels. The TDM concept is shown in Figure 4.28.

$$slotAllocated = (previousAllocatedSlot + 1)\%4 \qquad (4.1)$$

Contention-free routing is achieved with the help of different time slots. In contention-free routing, there is no collision between packets of multiple connections to access one output port. With the help of time slots, time to access the (Out) link is divided between multiple virtual channels. Every virtual channel equally shares the bandwidth of the (Out) link. The slots are allocated by the Scheduler of the router. Every router in the NoC increments one slot from the previous slot number, as shown in Figure 4.29. In the figure, slot 2 is dedicated to virtual channel B, while in the next router, the slot is incremented by 1 and slot 3 is assigned to virtual channel B. This helps to efficiently utilize the bandwidth, and one flit is routed from four routers in the consecutive four clock cycles. The "Scheduler" skips those slots that are not allocated to any of the virtual channels. This helps the connection to efficiently utilize the bandwidth of the interconnect and maintain the high throughput for that particular connection. The skipping of the unallocated time slots also increases the performance of a specific port, as it will dedicate the available bandwidth to already allocated active connections (Rijpkema et al. 2003).

If the bandwidth of the (Out) link physical channel is (bw) and there are (vc) virtual channels transmitting the flits (data), each virtual channel will have a throughput of TH_{min} (Gbps), as in Equation (4.2) (Kavaldjiev et al. 2006).

$$TH_{min} = \frac{bw}{vc} \qquad (4.2)$$

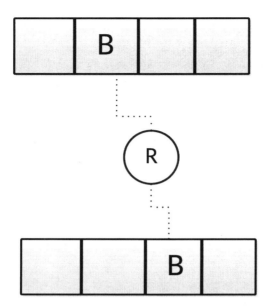

FIGURE 4.29 Router delays time slot allocation by one slot during connection setup.

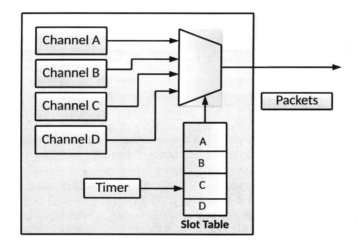

FIGURE 4.30 Scheduler architecture of a router with a slot table.

The bio-inspired NoC currently has four slots per port. So, there can be four virtual channels per port. The algorithm allocates the slots to the "Scheduler" so the throughput (TH_{min}) is guaranteed to each and every virtual channel (Kavaldjiev et al. 2006).

The Scheduler architecture of a router is shown in Figure 4.30. The slots are reserved during synapse establishment in the "OutPortCalc" module. Before deciding on any direction for the "synap" packet, the "OutPortCalc" checks whether any slot is available for this port. If yes, it will allocate the slot to the virtual connection;

Bio-Inspired NoC Fault-Tolerant Algorithms

otherwise, the "synap" is directed to another port based on the information present in the ISs. As shown, there are four virtual channels accessed by one router. The slot table is used to keep track of the slots allocated to each virtual channel.

The code in Figure 4.31 shows the allocation and checking of slots for "core-Port". If a time slot is available, only then can a virtual connection be established with "corePort". The slots are checked for all the cases mentioned in Figure 4.16, Figure 4.17, Figure 4.18 and Figure 4.19 before deciding on the direction of the "synap" packet. If (accessSlot==-1), then the slot (slotIndex) is not allocated to any connection. A total number of slots (numberOfSlot) at "Scheduler" are four.

There are four virtual channels (A, B, C, D), which are allocated to slots A, B, C and D. The timer variable synchronizes every slot at the "Scheduler". The "Scheduler" sends the packets (reqTDM) to read data from that port for a particular virtual channel. This helps with multiplexing and avoids contention between multiple connections.

In the code shown in Figure 4.32, the "Scheduler" is sending a "reqTDM" packet to a particular port to read a particular queue allocated to a certain slot (slotRead) and port (portIndex). This packet is sent on the "ctrl" port of the "Scheduler", as this is a control packet. In reply to this packet, "inPort" reads a particular queue allocated to this slot and sends the flit to a particular "Scheduler" of a port.

For every virtual channel, a separate queue is allocated at the "inPort" of the router, as shown in Figure 4.33. This helps avoid the HOL blocking problem when a fault occurs.

When the "inPort" reads a packet from the "Scheduler" to send the data, a particular queue is read from the "inPort", which is allocated to that slot. The "queue to read" parameter is provided by the "Scheduler". With the help of the TDM-based router, there is no collision between packets or deadlock scenario, as the "Scheduler"

```
for (int i=0;i<numberOfSlot;i++)
        {
                //Checking slots of current router of port 4
        accessSlot=routerSlots[routerIndex][4][slotIndex];
        if (accessSlot==-1)
          {
          //slot is free, allocate it to current port and make it 1
          routerSlots[routerIndex][4][slotIndex]=portIndex;
//Allocate this port and make its status as waiting for route reply so that it cannot send "request to send data" packet to
which it is allocated
                routerPortsStatusWaiting[routerIndex][4][slotIndex]=1;
          slotAllocated=slotIndex;
          swOutPortIdx = corePort;
          break;
          }
    }
```

FIGURE 4.31   Code for allocation and checking of slots.

```
NoCReqTDMMsg* reqTDM=new NoCReqTDMMsg("Request To Send
Data");
        reqTDM->setKind(NOC_REQTDM_MSG);
        reqTDM->setPortIndex(portIndex);
        reqTDM->setCurrentSlot(slotRead);
    send(reqTDM, "ctrl$o", outPort );
```

FIGURE 4.32   Code for sending reqTDM packet at "Scheduler".

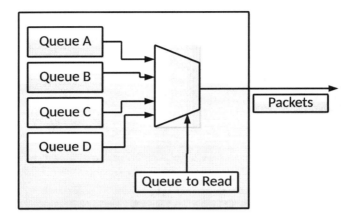

FIGURE 4.33 The inPort architecture of a router.

```
queueToRead=queueAllocatedToSlotPort[tdmRequestFromPort][tdmSlotFromPort];
NoCFlitMsg* tempFlit=(NoCFlitMsg*) Q[queueToRead].pop();
send(tempFlit, "out", outPort);
```

FIGURE 4.34 Code for reading a particular queue at inPort.

only allows one flit per router during one clock cycle. The code shown in Figure 4.34 specifies that "inPort" is receiving a request from "Scheduler" to read a particular queue. The "tdmRequestFromPort" and "tdmSlotFromPort" information is provided by the "Scheduler" when a particular slot is read. Later, the queue allocated to the particular slot is read. The flit that is saved in the queue is read and sent to the output port of the "inPort". This flit is received at the particular "Scheduler" of the port and "Scheduler" will send it to the (Out) link.

The flow of the packets in the bio-inspired NoC using the GT architecture is shown in Figure 4.35. The flow of packets is similar to the BE architecture except the use of a slot table at "Scheduler" and "Queues" at the "inPort". At every clock cycle, the slot is read at the "Scheduler". The "Scheduler" sends the "Request to send" packet to the port allocated in the slot. This packet is used to read a particular queue at the "inPort", which is allocated to this slot using a virtual channel. The communication between the source and destination proceeds like this until the end flit is sent by the source to terminate this connection.

4.8.3 Combined BE-GT Services in the NoC Architecture

To efficiently utilize the bandwidth of the NoC, BE services are provided along with GT services (Rijpkema et al. 2003). Figure 4.36 shows the combined BE-GT router port architecture. The packets with flits are received at the inPort, where the packet is saved either in the BE or GT queue as BQ and GQ, respectively. The source specifies the characteristics of the packet that is generated by the BE or GT source. The sources are selected using a random distribution formula. The slot is allocated to a BE- and

Bio-Inspired NoC Fault-Tolerant Algorithms

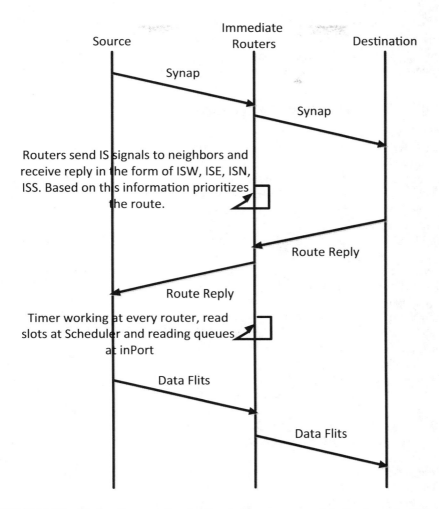

FIGURE 4.35 Timing diagram of packet flow between source and destination using GT.

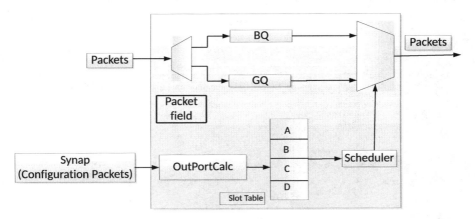

FIGURE 4.36 Port architecture for combined BE-GT services.

86   Bio-Inspired Fault-Tolerant Algorithms for Network-on-Chip

GT-based connection. The GT slot is permanent for the whole connection, while BE slot allocation will depend on the request from each packet containing flits. Besides that, the head flit of the BE packet will allocate one slot for the whole packet at the "Scheduler", while the end flit of that packet will clear that slot for other connections.

4.8.4 Bio-Inspired Algorithm for a BE- and GT-Based NoC

The bio-inspired algorithm is the same for both a GT- and BE-based NoC. In a GT-based NoC there is a slight change in the packet format of "synap" and "flit" compared to BE services due to the addition of queue and slot numbers, as shown in Figure 4.37 and Figure 4.38, respectively.

"Synapse ID" is unique for every connection between the source and destination. Whenever a fault occurs, the synapse ID is changed so that it is similar to the original synapse ID. This corresponds to the sprout behavior of the original synapse. "Source ID" and "Destination ID" correspond to the unique source PE and destination PE in the NoC. Output port numbers are saved in "Output port Numbers". This is used by the flits to send data on these ports. "Router IDs" are used by the route reply packet. The route reply packet is sent by the destination after the connection is established. "Slot Numbers" are saved in this field while the connection is established. These slot numbers are used by flits. "Queue Numbers" are also used by the flit to avoid HOL blocking. "Connection ID" corresponds to the number of connections a source can have; currently, four parallel connections are possible between a source and a destination. "Packet ID" corresponds to the packet number in a message. "Flit ID" is the flit number in a ten-flit packet. "Flit Type" corresponds to the header, body and tail flit. "Message Length" is the total number of flits in the packet.

In a GT-based NoC, with the help of time slots and queues at "Scheduler" and "inPort", multiple parallel connections are possible between multiple sources and multiple destinations. Figure 4.39 shows the case when multiple synapses are formed between the source (0) and destination (15). The source (0) initiated a connection to create a synapse with destination (15). As the ports are shared among multiple virtual channels, there can be multiple connections between one source and one destination. Connections between multiple sources and multiple destinations are also possible, as shown in Figure 4.40. Port 2 (south port) of router 6 is shared between

Synapse

Synapse ID	Source ID	Destination ID	Output Port Numbers	Router ID's	Slot Numbers	Queue Numbers	Connection ID

FIGURE 4.37   Synap packet format.

Flit

Packet ID	Flit ID	Connection ID	Flit Type	Message Length	Source ID	Destination ID	Slot Numbers	Queue Numbers	Output Port Numbers	Synapse ID

FIGURE 4.38   Flit packet format.

Bio-Inspired NoC Fault-Tolerant Algorithms 87

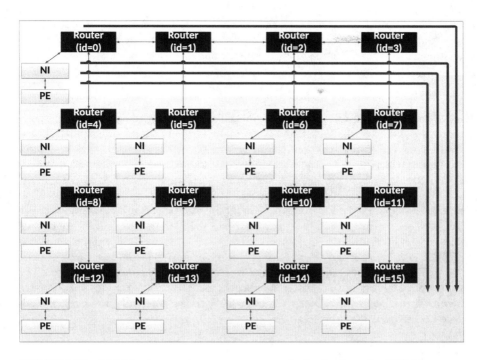

FIGURE 4.39 Multiple synapses between a source and a destination using TDM.

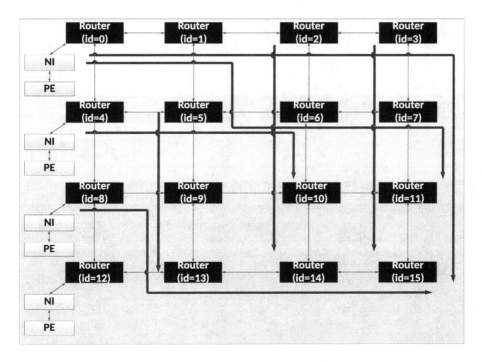

FIGURE 4.40 Multiple synapses between multiple sources and multiple destinations using TDM.

two connections. Two different slots are allocated to the connections. Similarly, ports 2 of router 7 and 9 are shared by two connections. The port 4 (core port) of router 0 and router 15 is shared among two and three connections, respectively.

SUMMARY

Bio-inspired algorithms bypass the path that has more faulty interconnects by prioritizing the path using IS packets information (two-hop information from their neighbors). If the interconnect or router is busy for a particular time, the various control packets and data flits are rescheduled to avoid deadlock and congestion. If a interconnect becomes faulty, a bio-inspired algorithm bypasses the complete row of routers, which helps avoid congestion. This makes the bio-inspired algorithm reliable and efficient. As there are no routing tables in bio-inspired algorithms, less on-chip area is taken up compared to routing table algorithms. But at the same time, the flow of control packets between neighbor routers makes bio-inspired algorithms somewhat complicated. Theoretically, bio-inspired NoC algorithms consume less power, as the "synap" packet moves in a directed path as controlled by the neighbor information. Bio-inspired algorithms are also implemented using BE, GT and combined BE-GT architectures, and it was found that the BE architecture efficiently utilized the bandwidth compared to GT services.

REFERENCES

Ben-Itzhak, Yaniv, Eitan Zahavi, Israel Cidon, and Avinoam Kolodny. 2012. "HNOCS: Modular open-source simulator for heterogeneous NoCs." *2012 International Conference on Embedded Computer Systems (SAMOS)*.

Breedlove, S Marc, Neil V Watson, and Mark R Rosenzweig. 2007. *Biological psychology: An introduction to behavioral and cognitive neuroscience*: Sinauer.

Hansson, Andreas, Kees Goossens, and Andrei Rădulescu. 2007. "Avoiding message-dependent deadlock in network-based systems on chip." *VLSI Design 2007*.

Hansson, Andreas, Mahesh Subburaman, and Kees Goossens. 2009. "Aelite: A flit-synchronous network on chip with composable and predictable services." *Proceedings of the Conference on Design, Automation and Test in Europe*.

Hashmi, Atif, Hugues Berry, Olivier Temam, and Mikko Lipasti. 2011. "Automatic abstraction and fault tolerance in cortical microarchitectures." *ACM SIGARCH Computer Architecture News*.

Kalat, James W. 2015. *Biological psychology*: Nelson Education.

Kavaldjiev, Nikolay, Gerard JM Smit, Pierre G Jansen, and Pascal T Wolkotte. 2006. "A virtual channel network-on-chip for GT and BE traffic." *IEEE Computer Society Annual Symposium on Emerging VLSI Technologies and Architectures (ISVLSI'06)*.

Nicopoulos, Chrysostomos, Vijaykrishnan Narayanan, and Chita R Das. 2009. *Network-on-chip architectures: A holistic design exploration*. Vol. 45: Springer Science & Business Media.

Rijpkema, Edwin, Kees Goossens, Andrei Rădulescu, John Dielissen, Jef van Meerbergen, Paul Wielage, and Erwin Waterlander. 2003. "Trade-offs in the design of a router with both guaranteed and best-effort services for networks on chip." *IEE Proceedings-Computers and Digital Techniques* 150 (5):294–302.

5 Analysis of Bio-Inspired NoC Fault-Tolerant Algorithms

5.1 RESEARCH FRAMEWORK, DESIGN AND PARAMETERS

The bio-inspired algorithm was tested with various faulty interconnects introduced during the simulation. The faults were introduced using the uniform random distribution formula. The seed value for the random distributed formula was changed in every iteration to make it randomized. To get the actual results without transient effects, the simulation was conducted thirty-five times for a 95% confidence interval for every fault. The injection rate of the packet with flits was changed from 10% to 100%. Injection rate is the percentage utilization of the interconnect bandwidth. The simulation was run multiple times based on the confidence interval measurement (Byrne 2013). The observed proportion (p) was chosen to be 90%, and the width of percentage error (w) to be 10%, while the $z_{\alpha/2}$ value was 1.96 for a 95% confidence interval. These values were put in the Equation (5.1) to measure the number of iterations ($n = 35$ turns).

$$n = p\left(1 - p\right)\left(\frac{z_{\alpha/2}}{w}\right) \tag{5.1}$$

Injection rate was linked to the bandwidth of the interconnect connected with the PE. In this simulation, the interconnect bandwidth was 16 Gbps. The injection rate could be varied by changing the number of idle cycles between the packets of wormhole switching. Equation 5.2 was used to calculate the number of idle cycles (Tedesco et al. 2005).

$$cycles = \left(\frac{CTR}{PETR}\right) * PSIZE * CyclesPerFlit \tag{5.2}$$

where *cycles* are the number of idle cycles between packets from the processing element (PE). CTR is the channel transmission rate in MBps (16,000 MBps) and PETR is the PE transmission rate (10% to 100%). PSize is packet size in terms of a number of flits (10 flits), while cycles per flit refer to the number of cycles required to transmit per flit in wormhole switching. Table 5.1 shows the number of idle cycles and delay time with reference to the injection rate (1 cycle = 2 ns).

To test the bio-inspired Network on Chip (NoC) fault-tolerant algorithm, a 4 × 4 NoC was used. The bio-inspired algorithms were tested for sixteen sources and

TABLE 5.1
Number of Idle Cycles vs. Injection Rate

Injection Rate	Delay Time (ns)	Number of Idle Cycles
100%	0	0
90%	2	1
80%	6	3
70%	8	4
60%	14	7
50%	20	10
40%	30	15
30%	46	23
20%	80	40
10%	180	90

sixteen destinations using a random distribution formula. The topology of the NoC was mesh. The clock frequency of the NoC was 500 MHz, the flit size was 4 bytes, and the packet length was ten flits. In each iteration, different sources were injecting various numbers of packets. A total of 100,000 packets or 1,000,000 flits were injected in the NoC to utilize the bandwidth of the NoC fully. The algorithm was tested using uniform and random traffic patterns. The simulation was run 50 ms per iteration. Wormhole switching was used in which the packet was divided into header, body and tail flits.

Bio-inspired algorithms calculated the quality of service (QoS) parameters of the NoC. The QoS is mainly linked with the latency and throughput parameters (Bolotin et al. 2004). Bio-inspired algorithms also analyzed other parameters of bandwidth utilization, latency, interflit arrival time and throughput utilization at the sink. As bio-inspired algorithms detected the faults on a per-router basis and the algorithm worked on a router-to-router basis, this made the bio-inspired algorithm more efficient at high traffic and fault rates (Patooghy, Miremadi, and Fazeli 2010).

An OMNET++–based heterogeneous Network on Chip simulator (HNOCS) (Ben-Itzhak et al. 2012), a state-of-the-art simulator, was used to implement the bio-inspired algorithms. It supports both synchronous and asynchronous communication. HNOCS is open source, and it supports all topologies. It supports parallelism and heterogeneous devices. The detailed comparison with other simulators is presented in the HNOCS paper cited earlier.

R-Software (Team 2013) was used to analyze the results statistically. With the help of the "sqldf" package of R-Software, multiple Structured Query Language (SQL) queries were executed to analyze the different results generated by HNOCS. The simulation framework is summarized in Table 5.2.

Analysis

TABLE 5.2
Simulation Framework

Function	Specification
Topology	4 × 4 mesh
Switching	Wormhole switching
Flow control techniques	Credit mechanism (BE), Nil (GT)
Services	BE, GT
Buffer requirement	Input buffer
Buffer size	16 packets
Packet size	10 flits
Flit size	4 bytes
Clock frequency	500 MHz
Routing algorithm	Bio-inspired routing algorithms, XY, odd-even
Traffic injection rate	Uniform and random
Tool	HNOCS
Performance evaluation	Bandwidth, latency, interflit arrival time, throughput, accepted traffic (flits/cycle/node)

5.2 BIO-INSPIRED NoC FAULT-TOLERANT ALGORITHMS ANALYSIS RESULTS

The synaptogenesis algorithm using best effort (BE) was implemented on faulty routers first and later on an interconnect basis. The algorithm was performing well on a faulty interconnect basis compared to a per–faulty router basis, as presented in Section 5.2.1.1. That is why all other bio-inspired algorithms were implemented using a faulty interconnect basis, and it was found that most of the techniques in the literature were also implemented on a faulty interconnect basis, as shown in Table 5.3 and Table 5.4. The synaptogenesis algorithm was improved by adopting the sprouting concept. The algorithms were implemented using BE communication service levels. The improved algorithm was applied on a faulty interconnect basis, and its performance was found to be better compared to the synaptogenesis algorithm. Subsequently, an improved bio-inspired algorithm was implemented using a guaranteed throughput (GT) architecture with a time division multiplexing (TDM) technique. Later, the combined BE-GT NoC architecture was implemented. The performance of the combined BE-GT architecture was found to be better than the overall BE and GT implementation, respectively, as presented in Sections 5.2.1.2 to 5.2.1.6.

5.2.1 PERFORMANCE EVALUATION PARAMETERS

The bandwidth was measured in terms of megabytes per second (MBps). This parameter corresponded to the bandwidth utilization of the NoC compared to the number of faults. Bandwidth corresponded to the total number of flits received at the destination in a particular time frame. The code used to compute the bandwidth is presented as follows:

```
noOfFlits++;
double BW_MBps = 1e-6 * noOfFlits * flitSize_B/(simTime().
dbl()-
timeAtWhichFirstFlitCreated);
bandwidth.record(BW_MBps);
```

Where:
noOfFlits is the total number of flits received at the destination.
BW_MBps is the measured bandwidth. It stores values to be saved in the vector array.
1e-6 is used to convert the values computed from the equation into mega notation.
flitSize_B is the flit size in byte (4 bytes).
simTime().dbl() is the current simulation time when the flit is received at the destination.
timeAtWhichFirstFlitCreated is the time when the first flit is created at the source.
bandwidth.**record**(BW_MBps) stores the value (BW_MBps) in the vector array.

Similarly, latency refers to the time taken by packets with flits to traverse from the source PE to destination PE. The code used to compute the latency of the flits is as follows:

```
double eed = (simTime().dbl()-msg->getCreationTime().
dbl());
double eed_ns = eed * 1e9;
end2EndLatencyVec.record(eed_ns);
```

Where:
eed is the end-to-end delay from the time the flit was created to the current simulation time. The current simulation time is the current time when the flit is received at the destination.
eed_ns is the time in nanoseconds.
end2EndLatencyVec.**record**(eed_ns) stores values (eed_ns) in the vector array.

The interflit arrival time was measured in ns. The interflit arrival time corresponded to the time at which the flits were received at the destination. This increased when the numbers of faults were increased as the network recovered from the faults. The following code was used at the destination:

Analysis

```
currentTime=simTime().dbl();
diffTime=currentTime-refernceTime;
referenceTime=currentTime;
timeFlitArrival.record(diffTime);
```

Where:

currentTime is the time when the flit is received at the destination.
referenceTime is the time when the previous flit was received. For the first flit, the creation time of the flit is chosen.
diffTime is the difference between currentTime and referenceTime.
timeFlitArrival.record(diffTime) stores values (diffTime) in a vector array.

The throughput utilization was measured at the destination PE. This parameter was affected by the arrival of flits at the destination. The code used to compute throughput is as follows:

```
currentTime=simTime().dbl();
diffTime=currentTime-referenceTime;
referenceTime=currentTime;
double throughput = (flitSize_B * 8)/diffTime;
throughput.record(throughput);
```

Where:

currentTime is the time when the flit is received at the destination.
referenceTime is the time when the previous flit was received at the destination. For the first flit, the creation time of the flit is chosen.
diffTime is the difference between currentTime and referenceTime.
double throughput calculates throughput for every flit received, divided by the time in which it is received.
throughput.record(throughput) records values (throughput) in the vector array.

The accepted traffic or saturation point is measured in terms of flit/cycle/node. This parameter corresponded to the point when the buffers of the NoC overflow and latency is increased to infinity. The code used to compute the accepted traffic is presented as follows:

```
double BW;
double acceptedTraffic = BW/clockTimePeriod * flitSize_B *
numberCycles
acceptedTraffic.record(acceptedTraffic);
```

Where:
BW is bandwidth in byte per second.
acceptedTraffic is measured in flit/cycle/node. It stores values to be saved in the vector array.
flitSize_B is the flit size in byte (4 bytes).
clockTimePeriod is equal to 0.5×10^9 cycles (1 cycle = 2 ns).
numberCycles is the number of clock cycles.
acceptedTraffic.**record**(acceptedTraffic) stores the value (acceptedTraffic) in the vector array.

Another way to compute the saturation point is by using the graphical method by plotting the accepted traffic rate vs. latency (cycles), as shown in Figure 5.11. The point on the x-axis where the latency increases to almost infinity is the saturation point.

5.2.1.1 Synaptogenesis Algorithm Using BE per Faulty Router and Interconnect Basis

Initially, the synaptogenesis algorithm was applied on a per-router faulty case. In this case, the complete router was blocked when it became faulty. Due to this, all five ports of the router were blocked, and they were not accessible for traversing the packet to the neighbor routers. Due to this, the number of routing possibilities is also reduced, as all five ports were blocked. This also reduces the performance of the NoC compared to the interconnect faulty case. Therefore, later, all bio-inspired algorithms in the literature were implemented on faulty interconnect basis. In this simulation scenario, the bio-inspired algorithm is only tested for one source and one destination at a time. The faults were introduced on a particular path to test the algorithm.

As can be seen in Figure 5.1, when there were no faulty routers and interconnects, the bandwidth utilization of the interconnect-based algorithm was better compared to the router-based bio-inspired algorithm, as the bandwidth utilization was increased from 1927 MBps to 1989 MBps. The overall bandwidth utilization increase was 5.03% from the router-based to interconnect-based bio-inspired routing algorithms. The reason is that in interconnect-based algorithms, there are more routing paths compared to the faulty router algorithm. Therefore, the bio-inspired algorithm can adapt any alternative path to bypass the fault.

Similarly, the latency of the interconnect-based bio-inspired algorithm was less compared to router-based bio-inspired algorithms. The reason was that as all the ports of the router are blocked and no routing was possible through complete routers, this increases the latency of the flits in the router-based algorithm, as shown in Figure 5.2. The latency of the flits in the interconnect-based algorithm reduced to almost 50 ns from 58 ns when three faults were introduced during the simulation. The overall latency decrease was 10.64% from router-based to interconnect-based bio-inspired routing algorithm.

Similarly, the interflit arrival time was reduced to 7.7 ns from 10 ns when three interconnects and routers became faulty in interconnect-based and router-based bio-inspired algorithms, respectively. The overall interflit arrival time decreased 17.25% from router-based to interconnect-based bio-inspired routing algorithm, as shown in

Analysis

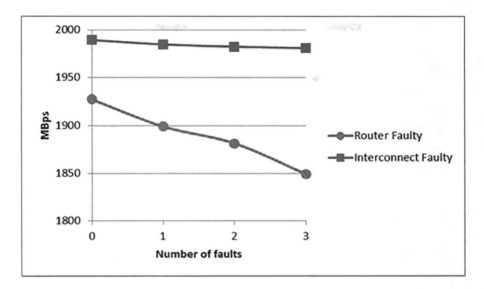

FIGURE 5.1 Bandwidth vs. number of faults.

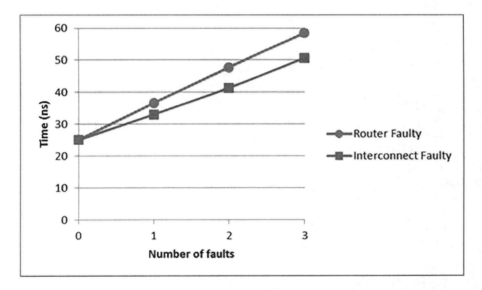

FIGURE 5.2 Latency vs. number of faults.

Figure 5.3. This reason was that the interflit arrival time should be more for complete router blockages compared to one interconnect blockage.

The graph in Figure 5.4 shows that the throughput of the improved interconnect-based bio-inspired algorithm also increases 13.75% from the previous router-based algorithm.

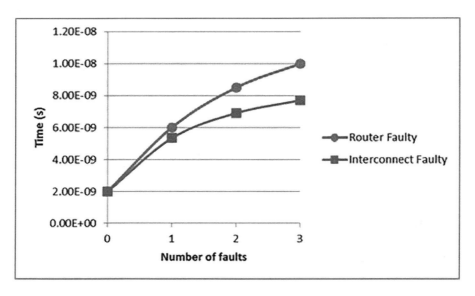

FIGURE 5.3 Interflit arrival time vs. number of faults.

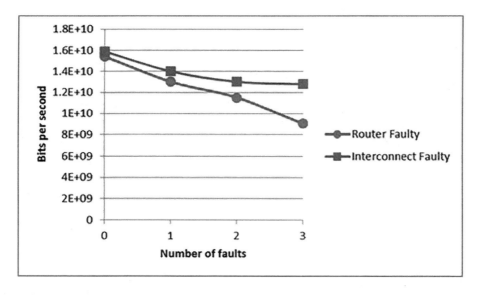

FIGURE 5.4 Throughput vs. number of faults.

To analyze the bio-inspired algorithms, they were thoroughly tested, validated and compared with the algorithms in the literature. For this reason, XY and odd-even routing algorithms were implemented. The results were then compared against the bandwidth, latency, interflit arrival time and throughput parameters. The bio-inspired algorithms have been found to perform better compared to the algorithms in the literature, as described in the following subsections.

Analysis

5.2.1.2 Bandwidth

The BE-based communication services complemented the GT services. At times, GT-based communication channels do not efficiently utilize the bandwidth, and this leads to underutilization of the resources. To efficiently utilize the bandwidth of the NoC, BE communication services were also provided along with the GT service level. The bandwidth, throughput, interflit arrival time and latency of BE and GT architectures depended on the traffic pattern and data flit injection rate. At a higher traffic injection rate, GT connections were preferred, as can be seen in Figure 5.5. At a 100% traffic injection rate, the bandwidth utilization of GT was found to better compared to the overall average TDM (bio-inspired TDM) performance for 10% to 100% injection rates for multiple faults, as there are more delay cycles between packets in the overall average TDM connections. The sprouting BE performed better compared to TDM-based GT connections. The sprouting BE efficiently utilized the bandwidth by 14.51% compared to 100% injection rate TDM-based connections, while the bandwidth utilization increased to 78.35% when the sprouting BE bandwidth utilization was compared with the bandwidth utilization of the average TDM connections. The sprouting BE performed better compared to TDM connections because of efficient bandwidth utilization compared to time slots allocation in GT-based TDM connections. The time slots divide the available bandwidth among multiple connections in GT algorithms, which at times leads to underutilization.

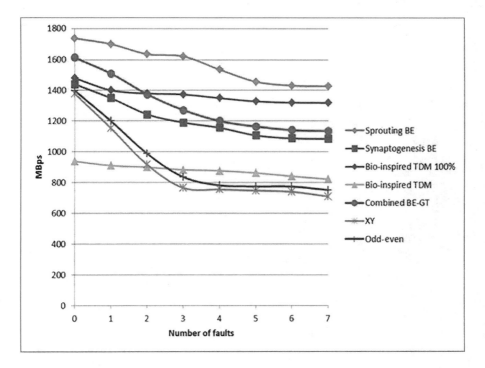

FIGURE 5.5 Bandwidth vs. number of faults.

The sprouting BE also performed better compared to the combined BE-GT algorithm by 20.50%, as GT connections at times underutilize the bandwidth. The sprouting BE efficiently utilized the bandwidth by 29.86% compared to the synaptogenesis BE. In the BE NoC, the synaptogenesis algorithm took 3 ns to detect the fault, while the sprouting BE quickly detected the fault and sprouts emerged from the neighbor router. The sprouting BE performed better compared to the algorithms in the literature of XY and odd-even by 74.99% and 67.07%, respectively. The fault detection mechanism in the bio-inspired algorithms was more efficient compared to the XY and odd-even algorithm, as bio-inspired algorithms had two-hop neighbor information. Furthermore, the restriction of turns in XY and odd-even further degraded their performance compared to bio-inspired algorithms. The odd-even routing algorithm performed better compared to XY by 4.75%, as it avoided the circular wait at the NoC due to restrictions in turns.

The overall performance of BE-based algorithms was better compared to overall TDM-based GT connections by 23.44% in terms of bandwidth utilization. Similarly, the combined BE-GT algorithm performed better compared to overall BE- and GT-based algorithms by 6.65% and 15.75%, respectively.

5.2.1.3 Latency

The sprouting BE had the least end-to-end latency from source PE to destination PE compared to the literature and other bio-inspired algorithms, as shown in Figure 5.6. As the number of faults increased, the latency of the synaptogenesis BE was higher

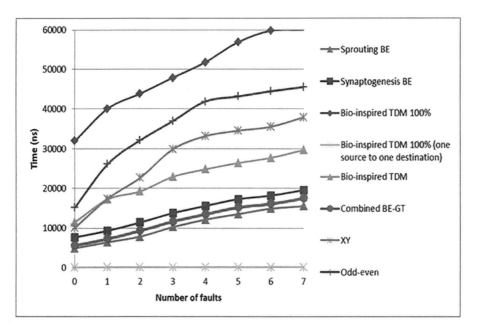

FIGURE 5.6 Latency vs. number of faults.

Analysis

than the sprouting BE by 31.95%, as it took more numbers of cycles (3 ns) to detect the fault and initiate a new synapse connection. Similarly, the latency of the combined BE-GT was higher compared to the sprouting BE by 13.22%, as GT connections at times have more latency due to time slot allocations. The time slots with the GT were allocated to every connection, which degraded the performance of the NoC, as the bandwidth and time period were reserved for every connection. The 100% TDM connections had the highest latency when all sources injected traffic at the highest rate; this then led to the buffer overflows, and the latency of packets increased significantly, as shown in the figure. However, 100% TDM connections were also tested with one source and one destination scenario, as seen in the figure, and this had the least latency, as the time slots and bandwidth were not reserved for other connections. The magnified graph of the bio-inspired 100% TDM for one source and one destination is shown in Figure 5.7.

The average TDM connections (bio-inspired TDM) had 109.98% more latency as compared to the sprouting BE because of the time slot allocation. Another reason for the average TDM connections was more delay cycles between packets at low injection rates, which increased the latency of the packets. Similarly, the percentage increase in latency was 158.82% and 234.05% for XY and odd-even compared to the sprouting BE, respectively. The latency of flits increased in XY and odd-even because these algorithms lacked adaptivity. The latency of the odd-even algorithm was more compared to XY by 29.10% due to restrictions on certain turns in the NoC. Initially, at zero faults, the latency of the XY is less compared to the bio-inspired TDM because of the deterministic nature of XY, but as the number of faults increased, XY latency increased, as it lacks the adaptiveness compared to the bio-inspired TDM.

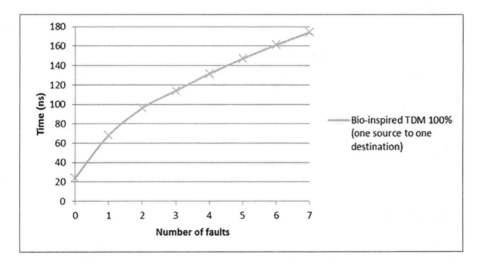

FIGURE 5.7 Enlarged plot of latency vs. number of faults for bio-inspired 100% TDM (one source to one destination).

The overall latency of GT-based algorithms was more compared to the overall BE-based algorithms by 188.40% because of the time slot allocation. Similarly, the overall BE- and GT-based algorithms had more latency compared to the combined BE-GT algorithm by 2.43% and 195.40%, respectively, as the combined BE-GT had the benefits of both packet switching (BE) and a connection-oriented mechanism (GT).

5.2.1.4 Interflit Arrival Time

Figure 5.8 shows the interflit arrival time for different fault-tolerant algorithms vs. the number of faults. The sprouting BE had the least interflit arrival time compared to the literature and other bio-inspired algorithms. At a 100% traffic injection rate, the interflit arrival time of the TDM was more compared to the sprouting BE by 21.56%. Similarly, the average TDM (bio-inspired TDM) connections had a larger interflit arrival time by 178.37% compared to the sprouting BE algorithm. The GT connections had a larger interflit arrival time because of the time slot allocation. The average TDM connections had the highest interflit arrival time because at a low injection rate there are more delay cycles between packets, as shown in Table 5.1. The interflit arrival time for the average TDM from 100% to 50% was also plotted, which shows that it was 35.87% more compared to the sprouting BE. However, the interflit arrival time increased as the injection rate was reduced, which in turn increased the delay cycles between packets. The combined BE-GT connections had more interflit

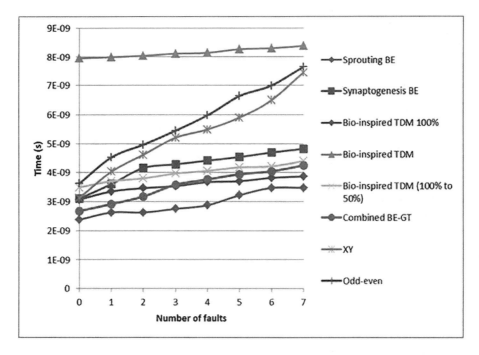

FIGURE 5.8 Interflit arrival time vs. number of faults.

Analysis

arrival time compared to the sprouting BE by 21.06% because of the time slot allocation. The synaptogenesis BE had a larger interflit arrival time by 43.61% compared to the sprouting BE. The synaptogenesis BE takes 3 ns to detect a fault compared to the sprouting BE. Similarly, the sprouting BE performed better compared to the algorithms in the literature of XY and odd-even by 80.63% and 95.56%, respectively, because of the two-hop neighbor information. The interflit arrival time of Odd-Even algorithm was more compared to XY by 8.27%. The interflit arrival time of XY was less compared to the bio-inspired TDM (100% to 50%) algorithm at zero faults, as it is a deterministic algorithm, which lacks the adaptiveness and follows a fixed path. However, as the number of faults increased, the interflit arrival time of the XY was larger compared to the bio-inspired TDM (100% to 50%) algorithm, as it lacks the efficient fault detection mechanism compared to the bio-inspired algorithm.

The interflit arrival time of the overall GT-based algorithms is larger compared to the overall BE-based algorithms by 64.17% due to the time slot mechanism. Similarly, the overall BE- and GT-based algorithms had a larger interflit arrival time compared to the combined BE-GT algorithm by 0.62% and 65.19%, respectively.

5.2.1.5 Throughput

The throughput of the sprouting BE was higher compared to the literature and other bio-inspired fault-tolerant algorithms, as shown in Figure 5.9. The sprouting BE throughput was higher compared to the 100% TDM and average TDM connections by 11.71% and 30.36%, respectively, as time slot allocation in the TDM leads to

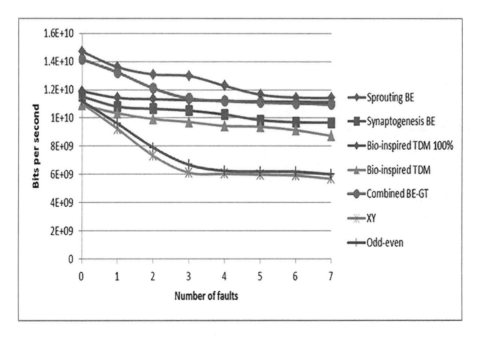

FIGURE 5.9 Throughput vs. number of faults.

underutilization of the resources. The sprouting BE performed better compared to the synaptogenesis BE and combined BE-GT in terms of throughput by 21.92% and 6.29%, respectively. The synaptogenesis algorithm takes more time to detect a fault compared to the sprouting algorithm. The combined BE-GT algorithm throughput was more compared to the 100% TDM; however, as the number of faults was increased, the complexity of the combined BE-GT algorithm increased, which lead to a drop in throughput. As can be seen in the figure, the throughput of the combined BE-GT is almost equal to the 100% TDM after three faults. The throughput of the sprouting BE was better compared to the algorithms in the literature of XY and odd-even by 76.41% and 68.42%, respectively, as it had two-hop neighbor information compared to the XY and odd-even.

The overall performance of the BE-based algorithms were found to be better compared to the overall GT-based algorithms by 9.50%. Similarly, the combined BE-GT algorithm efficiently utilized the throughput compared to the overall BE- and GT-based algorithms by 3.37% and 13.19%, respectively.

The sprouting BE algorithm was also tested for one source and one destination to see the robustness of the algorithm and quick recovery from faults, as shown in Figure 5.10. The figure shows the throughput of the NoC when various interconnects were made faulty. The throughput decreased for a very short period of time (in nanoseconds) as the network found the newer workaround synapse when a fault occurred. The flits traversed on the older synapse (the main connection between the source and destination), which is why the throughput degraded slightly and quickly recovered the maximum data rate (16 Gbps). The throughput was reduced by 37.22% during recovery from faults. The throughput dropped to 4 Gbps when the first fault occurred during the communication. Then, it was dropped to 3.2 Gbps, 2.67 Gbps, 4 Gbps, 3.2 Gbps, 2.67 Gbps, 4 Gbps and 3.2 Gbps for two, three, four, five, six, seven and eight

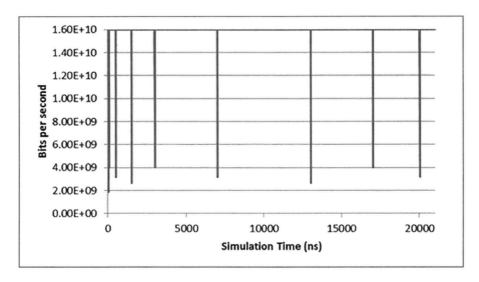

FIGURE 5.10 Sprouting BE throughput vs. number of faults.

Analysis

faulty interconnects, respectively, during the simulation. The average recovery time from faults and reception of flits at the destination was 9.71 ns. After this time, the NoC quickly regained the maximum throughput of 16 Gbps.

5.2.1.6 Accepted Traffic (Flit/Cycle/Node) or Saturation Point

Table 5.3 and Table 5.4 show the saturation point or accepted traffic (flit/cycle/node) of different fault-tolerant algorithms, where "–" shows that values are not mentioned in their paper. The accepted traffic of the bio-inspired algorithm with combined BE-GT services was found to be better than that of our previous work by 17.14%,

TABLE 5.3

Fault-Tolerant Algorithms Comparison (4 × 4 NoC)

Algorithms	Accepted Traffic(flits/cycle/node)							
	Number of Faults	0	1	2	3	4	5	6	7
Bio-inspired algorithm with combined BE-GT services (Sethi, Hussin, and Hamid 2018)	0.41	0.38	0.34	0.32	0.30	0.29	0.29	0.28
Bio-inspired algorithm using GT (Sethi, Hussin, and Hamid 2016)	0.52	0.32	0.28	0.25	0.22	0.2	0.18	0.17
Synaptogenesis algorithm using BE (Sethi, Hussin, and Hamid 2013b, 2014)	0.39	0.31	0.29	0.28	0.27	0.26	0.25	0.24
Sprouting algorithm using BE (Sethi, Hussin, and Hamid 2013a, 2014)	0.42	0.40	0.34	0.30	0.22	0.21	0.19	0.18
DBP (Koibuchi et al. 2008)	0.22	0.18	0.17	0.16	0.15	0.15	0.14	0.14
ODD-EVEN (Castro and de Lima 2013)	–	0.4	–	0.4	–	0.4	–	–
Inverted ODD-EVEN (Castro and de Lima 2013)	–	0.4	–	0.4	–	0.4	–	–
FTCAR (Valinataj et al. 2010)	0.31	–	–	0.25	–	0.24	–	–
HiPFaR (Ebrahimi, Daneshtalab, and Plosila 2013)	0.37	0.35	–	–	–	–	–	–

(Continued)

TABLE 5.3 (Continued)

Algorithms	Accepted Traffic(flits/cycle/node)							
	Number of Faults	0	1	2	3	4	5	6	7
Baseline (Ebrahimi, Daneshtalab, and Plosila 2013)		0.38	0.32	–	–	–	–	–	–
RAFT2 algorithm using VOPD traffic (Valinataj 2011)		0.26	–	0.23	–	–	–	–	–
RAFT2 algorithm using uniform traffic (Valinataj 2011)		0.37	–	0.29	–	–	–	–	–
Fault-Aware Dynamic Routing Algorithm (Hosseini, Ragheb, and Massoud 2008)		0.2	0.2	0.2	–	–	–	–	–
Flooding Algorithm (Hosseini, Ragheb, and Massoud 2008)		0.10	0.10	0.10	–	–	–	–	–

TABLE 5.4
Fault-Tolerant Algorithms Comparison (With Different Network Sizes)

Algorithms	Accepted Traffic(flits/cycle/node)								
	Number of Faults / NoC Size	0	1	2	3	4	9	16	20
Bio-inspired algorithm with combined BE-GT services (Sethi, Hussin, and Hamid 2018)	4×4	0.41	0.38	0.34	0.32	0.30	–	–	–
Bio-inspired algorithm using GT (Sethi, Hussin, and Hamid 2016)	4×4	0.52	0.32	0.28	0.25	0.22	–	–	–
Synaptogenesis algorithm using BE (Sethi, Hussin, and Hamid 2013b, 2014)	4×4	0.39	0.31	0.29	0.28	0.27	–	–	–
Sprouting algorithm using BE (Sethi, Hussin, and Hamid 2013a, 2014)	4×4	0.42	0.40	0.34	0.30	0.22	–	–	–
Typical Virtual Channel Management (Latif et al. 2011)	5×5	0.18	–	0.16	–	–	–	–	–

(Continued)

Analysis

TABLE 5.4 (Continued)

Algorithms	Accepted Traffic(flits/cycle/node)								
	Number of Faults / NoC Size	0	1	2	3	4	9	16	20
PVS (Latif et al. 2011)	5 × 5	0.22	–	0.20	–	–	–	–	–
FVS (Latif et al. 2011)	5 × 5	0.23	–	–	–	–	–	–	–
A Reconfigurable Routing Algorithm (Zhang, Greiner, and Taktak 2008)	5 × 5	0.28	0.27	0.22	0.16	0.15	–	–	–
RAFT (Valinataj, Liljeberg, and Plosila 2013)	5 × 5	–	–	–	–	–	0.28	–	–
Agent-based routing (Valinataj, Liljeberg, and Plosila 2013)	5 × 5	–	–	–	–	–	0.3	–	–
XY (Zhang, Han, and Zhang 2013)	8 × 8	0.36	–	–	–	–	–	0.25	–
FFAR (Zhang, Han, and Zhang 2013)	8 × 8	0.35	–	–	–	–	–	0.28	–
OE (Pasricha et al. 2010)	9 × 9	–	0.04	–	–	–	–	–	0.036
IOE (Pasricha et al. 2010)	9 × 9	–	0.035	–	–	–	–	–	0.03
OE+IOE (Pasricha et al. 2010)	9 × 9	–	0.06	–	–	–	–	–	0.065

while compared to the algorithms in the literature, it was better by 38.99%. The bio-inspired algorithm using GT had the best saturation point of 0.52 at zero faults compared to the fault-tolerant algorithms in the literature that show the efficiency and simplicity of the algorithm. Figure 5.11 shows the saturation point of the GT-based NoC as the number of faults is increased. This also demonstrates that the bio-inspired algorithm quickly found the path from the source PE to destination PE when there was no fault (1 cycle = 2 ns).

Most of the fault-tolerant algorithms have been found to be not thoroughly tested using different numbers of faults compared to the bio-inspired algorithm. The bio-inspired algorithms were also compared with the DBP (Koibuchi et al. 2008) fault-tolerant algorithm that is a thoroughly analyzed, tested and presented algorithm in the literature. The bio-inspired algorithm with combined BE-GT performed better compared to the DBP fault-tolerant algorithm by 99.24%. Similarly, the bio-inspired algorithm using GT and the bio-inspired algorithm using BE (synaptogenesis and sprouting) performed better by 63.36% and 72.12%, respectively, compared to the DBP fault-tolerant algorithm.

The accepted traffic rate of the bio-inspired NoC using GT was less compared to the BE-based bio-inspired algorithms. This was expected, as these algorithms

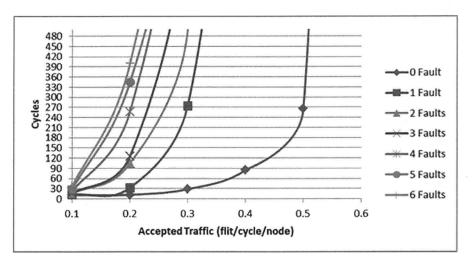

FIGURE 5.11 Accepted traffic for various numbers of faults.

are based on the BE scenario. In BE, the bandwidth and throughput utilization were found to be better compared to GT-based algorithms. The efficient utilization of the TDM-based GT connections depended on the burst of traffic from the source or from neighbor routers. Usually, TDM-based virtual circuits are allocated to those PEs that need guaranteed and faster services, while BE algorithms are packet-switching techniques that utilize all the available bandwidth without any reservation. The saturation point of the algorithms mentioned in (Castro and de Lima 2013) was slightly higher in three faulty cases (one, three and five). The saturation point of 0.4 for three different faulty cases was not a common behavior in the literature. Usually, the saturation point decreases as the number of faults is increased. The network saturates early, as the network has fewer paths for packets to traverse through and these paths are occupied by all the PEs available on the NoC. For this reason, the NoC saturates early, and it leads to overflow of buffers and queues. Moreover, the algorithms (Castro and de Lima 2013; Ebrahimi, Daneshtalab, and Plosila 2013; Valinataj 2011) (Hosseini, Ragheb, and Massoud 2008; Latif et al. 2011; Zhang, Greiner, and Taktak 2008; Valinataj, Liljeberg, and Plosila 2013; Zhang, Han, and Zhang 2013; Pasricha et al. 2010) were only tested with a traffic injection rate from 10% to 50%, while the bio-inspired algorithms were tested with a 10% to 100% injection rate. The bio-inspired NoC-accepted traffic is better than the fault-tolerant algorithms (Valinataj et al. 2010) for different faulty cases.

SUMMARY

The bio-inspired algorithms with BE, GT and combined BE-GT services performed better compared to the algorithms in the literature and had better throughput and bandwidth utilization. The overall BE-based algorithms efficiently utilized

Analysis

the bandwidth and maximized the throughput compared to GT-based algorithms. Similarly, the interflit arrival time and latency of the overall BE-based algorithms were less compared to the GT-based algorithms. The results showed that the bio-inspired NoC algorithm using combined BE-GT communication services performed better compared to fault-tolerant algorithms in the literature.

REFERENCES

Ben-Itzhak, Yaniv, Eitan Zahavi, Israel Cidon, and Avinoam Kolodny. 2012. "HNOCS: Modular open-source simulator for heterogeneous NoCs." *2012 International Conference on Embedded Computer Systems (SAMOS).*

Bolotin, Evgeny, Israel Cidon, Ran Ginosar, and Avinoam Kolodny. 2004. "Cost considerations in network on chip." *INTEGRATION, the VLSI Journal* 38 (1):19–42.

Byrne, Michael D. 2013. "How many times should a stochastic model be run? An approach based on confidence intervals." *Proceedings of the 12th International Conference on Cognitive Modeling*, Ottawa.

Castro, Helano S, and Otavio Alcantara de Lima. 2013. "A fault tolerant NoC architecture based upon external router backup paths." *2013 IEEE 11th International New Circuits and Systems Conference (NEWCAS).*

Ebrahimi, Masoumeh, Masoud Daneshtalab, and Juha Plosila. 2013. "High performance fault-tolerant routing algorithm for NoC-based many-core systems." *2013 21st EUROMICRO International Conference on Parallel, Distributed, and Network-Based Processing.*

Hosseini, Amir, Tamer Ragheb, and Yehia Massoud. 2008. "A fault-aware dynamic routing algorithm for on-chip networks." *2008 IEEE International Symposium on Circuits and Systems.*

Koibuchi, Michihiro, Hiroki Matsutani, Hideharu Amano, and Timothy Mark Pinkston. 2008. "A lightweight fault-tolerant mechanism for network-on-chip." *Proceedings of the Second ACM/IEEE International Symposium on Networks-on-Chip.*

Latif, Khalid, Amir-Mohammad Rahmani, Kameswar Rao Vaddina, Tiberiu Seceleanu, Pasi Liljeberg, and Hannu Tenhunen. 2011. "Enhancing performance sustainability of fault tolerant routing algorithms in NoC-based architectures." *2011 14th EUROMICRO Conference on Digital System Design.*

Pasricha, Sudeep, Yong Zou, Dan Connors, and Howard Jay Siegel. 2010. "OE+ IOE: A novel turn model based fault tolerant routing scheme for networks-on-chip." *2010 IEEE/ACM/ IFIP International Conference on Hardware/Software Codesign and System Synthesis (CODES+ ISSS).*

Patooghy, Ahmad, Seyed Ghassem Miremadi, and Mahdi Fazeli. 2010. "A low-overhead and reliable switch architecture for Network-on-Chips." *Integration* 43 (3):268–278.

Sethi, Muhammad Athar Javed, Fawnizu Azmadi Hussin, and Nor Hisham Hamid. 2013a. "Implementation of biological sprouting algorithm for NoC fault tolerance." *2013 IEEE International Conference on Circuits and Systems (ICCAS).*

Sethi, Muhammad Athar Javed, Fawnizu Azmadi Hussin, and Nor Hisham Hamid. 2013b. "Synaptogenesis based bio-inspired NoC fault tolerant interconnects." 2013 IEEE *International Conference on Control System, Computing and Engineering.*

Sethi, Muhammad Athar Javed, Fawnizu Azmadi Hussin, and Nor Hisham Hamid. 2014. "Bio-inspired NoC fault tolerant techniques." *2014 5th International Conference on Intelligent and Advanced Systems (ICIAS).*

Sethi, Muhammad Athar Javed, Fawnizu Azmadi Hussin, and Nor Hisham Hamid. 2016. "Biologically inspired network on chip fault tolerant algorithm using time division multiplexing." *2016 6th International Conference on Intelligent and Advanced Systems (ICIAS).*

Sethi, Muhammad Athar Javed, Fawnizu Azmadi Hussin, and Nor Hisham Hamid. 2018. "Bio-inspired network on chip having both guaranteed throughput and best effort services using fault-tolerant algorithm." *IEEJ Transactions on Electrical and Electronic Engineering* 13 (8):1153–1162.

Team, R Core. 2013. "R: A language and environment for statistical computing."

Tedesco, Leonel, Aline Mello, Diego Garibotti, Ney Calazans, and Fernando Moraes. 2005. "Traffic generation and performance evaluation for mesh-based NoCs." *Proceedings of the 18th Annual Symposium on Integrated Circuits and System Design.*

Valinataj, Mojtaba. 2011. "Evaluation of fault-tolerant routing methods for NoC architectures." *2011 14th EUROMICRO Conference on Digital System Design.*

Valinataj, Mojtaba, Pasi Liljeberg, and Juha Plosila. 2013. "Enhanced fault-tolerant Network-on-Chip architecture using hierarchical agents." *2013 IEEE 16th International Symposium on Design and Diagnostics of Electronic Circuits & Systems (DDECS).*

Valinataj, Mojtaba, Siamak Mohammadi, Juha Plosila, and Pasi Liljeberg. 2010. "A fault-tolerant and congestion-aware routing algorithm for networks-on-chip." *13th IEEE Symposium on Design and Diagnostics of Electronic Circuits and Systems.*

Zhang, Shijian, Guodong Han, and Fan Zhang. 2013. "Very fine-grained fault-tolerant routing algorithm of NoC based on buffer reuse." *2013 IEEE 4th International Conference on Software Engineering and Service Science.*

Zhang, Zhen, Alain Greiner, and Sami Taktak. 2008. "A reconfigurable routing algorithm for a fault-tolerant 2D-mesh network-on-chip." *2008 45th ACM/IEEE Design Automation Conference.*

6 Conclusion and Future Work

Network on Chip (NoC) refers to a communication standard for on-chip networks. NoC has replaced the traditional bus and crossbar interconnection of System on Chip (SoC) components. Multiprocessor SoC (MPSoC) is also unable to cope with the globally asynchronous locally synchronous (GALS) communication. Therefore, this leads to the concept of an NoC that can provide the modularity, scalability and efficient reuse of the resources with high bandwidth.

The excessive and parallel communication requirements of heterogeneous processing elements (PEs) in the NoC have made the communication structure very complicated. The size of the devices is scaled down to support the complexity, which has contributed to the faults. Fault tolerance is becoming a very important concept as more devices are favoring NoC. Fault tolerance separates the data communications from the on-chip communication mechanism. Fault tolerance is the way to cope with both permanent and temporary faults. Permanent faults are due to the physical damage to resources on the NoC, such as electromigration, die-electric breakdown, manufacturing defects or any physical damage. They can be addressed with the help of redundant routers, interconnects or resources (PEs). Temporary faults are due to changes in voltage, temperature, congestion in routers or faults. These faults can be addressed with fault-tolerant routing algorithms.

Different fault-tolerant algorithms have been proposed. Deterministic, stochastic, fully adaptive and partial adaptive routing algorithms are four broad categories of fault-tolerant algorithms. However, all of these algorithms do not entirely address the drawbacks. Traditional fault-tolerant algorithms also suffer from adaptivity and robustness limitations. Biological brain-inspired fault-tolerant algorithms have been proposed, making the NoC more fault-tolerant and robust. Synaptogenesis and sprouting are two of these biological brain-inspired algorithms. The bio-inspired algorithm "synaptogenesis" solves the problem of dynamic interconnections between various PEs by adapting the mechanism of the growth cone present at the top of the axon terminal and the dendrites. The "sprouting" algorithm helps to recover from faults by generating a sprout from the neighbor router. The bio-inspired algorithm improves the overall performance of the NoC by rerouting the packets to the newly established synapse to bypass the faulty interconnects.

In the bio-inspired algorithm, the "synap" packet is generated by source PE to have a connection with the destination PE. Upon reception of the "synap" packet at the router, the router sends information set (IS) packets to neighbor routers so they have two-hop information. The neighbor routers reply with IS response packets. After receiving IS response packets and based on the destination PE, the router prioritizes the path toward the destination. After traversing the routers, the "synap"

109

signal is reached at the destination. The destination PE sends a route reply signal to the source PE. Data flits traverse over this connection until there are no faults. When interconnects become faulty during the communication, the neighbor router detects the fault and initiates the synapse connection (sprout) between the current router to the older synapse or to the destination router. If there is no older synapse to connect with, the sprout is directly connected with the destination PE. The connection with the older synapse avoids the traversal of unnecessary routers—this makes the algorithm robust and fault-tolerant.

The bio-inspired algorithm was implemented via three communication setups; those are guaranteed throughput (GT), best effort (BE) and combined BE-GT using three different NoC architectures. There are basically two broad categories of quality of services parameters in NoC, namely GT service and BE service. In GT, the resources on the network are reserved for a particular time period between source and destination PEs. The circuit-switching or connection-oriented mechanism of packet switching is mainly used to allocate resources and guarantee throughput. GT connections are usually allocated to real-time traffic because of high traffic and less latency requirements. GT connections address the worst-case scenario because the resources can be silent for a particular time, which makes the bandwidth utilization low at times. BE services complement the GT connections because of their underutilization of the channel bandwidth. BE is the packet-switching concept that utilizes the unallocated bandwidth to sources and other routers. BE traffic does not allocate any resources on the NoC. The BE service represents the average-case scenario. The channel is always utilized when the source and destination need it. Video processing and multimedia applications are examples of GT critical traffic, while cache updates are examples of BE noncritical traffic.

BE services were provided using packet-switching techniques that have credit-based flow control. In credit-based flow control, there was a credit counter at the router, which controls the flow of flits between routers. If the credit counter is zero, then this port is busy and not allowed to send more flits until the neighbor router replies with the credit increment packet after receiving the flit. The GT service was provided using time division multiplexing (TDM). In TDM, the output link bandwidth was divided using time period slots and allocated to multiple virtual channels. To avoid the head of line (HOL) blocking, multiple buffers were allocated at the input port. The HOL problem occurred due to the blockage of header flits. The multiple buffers at the input port were allocated to each virtual channel. In the GT service, no flow control technique was required, as the resources were reserved for a particular time period. The results showed that the bio-inspired NoC algorithm using BE, GT and combined BE-GT communication services performed better compared to the fault-tolerant algorithms found in the literature.

6.1 FUTURE WORK

The future work should include the implementation of a complete bio-inspired NoC in a field programmable gate array (FPGA) and further optimize the algorithm and delay the saturation point. Another possible future project would be to update the

Conclusion and Future Work

routing algorithm so that it can avoid more faulty and congested paths in the NoC by processing the information received from neighbor routers, where the information is received in the form of ISW, ISE, ISS and ISN packets. Moreover, future work should also include further testing of the bio-inspired algorithm on different NoC sizes and topologies.

Appendix A

Survey of NoC Architectures: Study 1

S. No.	Architectures	Comparison Parameters												
		Link Sharing	Routing	QoS	Switching Techniques	Evaluation Parameters	NoC Clocking Mechanism	Topology Supported (2D)	Open Source	Connection Types	Implementation	NoC Size (for experiments)	Area	Power Dissipation/ Energy Consumption
1	SPIN (Guerrier and Greiner 2008) (Scalable Programmable Interconnect Network)	Packet switching (round-robin arbitration)	Adaptive and distributed	BE	Wormhole switching using credit mechanism	—Area —Message latency —Network throughput —Number of cycles vs. number of cores	Synchronous	Fat Tree	–	1–1, broadcast	Simulation & Synthesis (Cycle Accurate System Simulator, CASS)	—256 terminals —10X10 partial crossbar	–	–
2	aSOC (Liang, Swaminathan, and Tessier 2000) (Adaptive System on Chip)	Packet switching (TDM)	Static	GS	Header	—Time —Link utilization —Used link —Area —Frequency	Synchronous	Mesh	–	1–1 (Point-to-point communication)	—Simulation (NSIM Parallel Interconnect and SimpleScalar) & —Synthesis (Altera Quartus)	3X3 4X4 Mesh	—Router area is 6000X5000λ^2 —ASIC implementation of switch have 50,000 transistors	–
3	Dally et al. (Dally and Towles 2001)	Circuit switching	Deterministic	Packet	Wormhole switching & header	—Area —Power	Synchronous	Folded torus	–	–	–	–	0.59mm^2	–
4	MicroNetwork (Wingard 2001)	Packet switching (TDM Using resource arbitration scheme)	–	Packet	–	–	–	Custom (bus-based)	–	–	Simulator	–	–	

5	OCTAGON (Karim, Nguyen, and Dey 2002)	Packet & circuit switching	Deterministic routing	GS	Wormhole switching	—Latency vs. utilization —Packet loss probability vs. egress queue size —Nodes vs. links vs. hops Number of links vs. number of nodes	Nodes in ring of degree 3	Octagon	–	1–1	Simulator	8 node	140000 number of transistors are required for VC=4	Switch consumes 20955mW
7	CHAIN (Bainbridge and Furber 2002	Packet switching	Source	Packet	Header with separate control lines (EOP)	—Area —Throughput —Latency	Asynchronous	Any	–	1–1, multicast	Smart card chip (ASIC)	—Two links for forward and reverse communication —Multiplexer and demultiplexer topology —Initiator and Target —Command switch —Response switch	115,000 transistors	–
6	PROTEO (Siguenza-Tortosa and Nurmi 2002)	Packet switching	Adaptive	Packet	Header	–	Globally Asynchronous Locally Synchronous (GALS) or Asynchronous	Hierarchical	–	–	Simulation & Synthesis (VHDL)	Having ring along with star or bus topology	–	–

(Continued)

S. No.	Architectures	Comparison Parameters												
		Link Sharing	Routing	QoS	Switching Techniques	Evaluation Parameters	NoC Clocking Mechanism	Topology Supported (2D)	Open Source	Connection Types	Implementation	NoC Size (for experiments)	Area	Power Dissipation/ Energy Consumption
8	CLICHÉ (Kumar et al. 2002) (Chip Level Integration of Communication Heterogeneous Elements)	Packet switching	Deterministic	Packet	Wormhole switching	—Performance —Utilization —Capacity —Wireability —Delay —Bandwidth —Position of pads & repeaters	Globally Asynchronous Locally Synchronous (GALS)	Mesh	–	Multipath	Simulation (NS-2)	5X5 Mesh	–	–
9	RAW (Taylor et al. 2002)	Packet switching	Dimension order	Packet	Wormhole switching	–	–	Mesh	–	–	IBM SA-27E (ASIC)	1024 tiles	122 million transistors	25W
10	Eclipse (Forsell 2002)	Packet switching	Deterministic	–	Store & forward	–	–	Mesh	–	–	Simulator	–	–	–
11	Æthereal (Rijpkema et al. 2003)	Packet switching using TDM (round-robin for BE & no arbitration for GT)	Source along with contention free	GS, BE (Distributed routing with BE)	Wormhole switching	—Area —Latency —Throughput —Frequency —Switch size —Service Class —Programming model	Globally Synchronous with mesochronous or asynchronous links	Mesh	No	1–1, multicast	Synthesis	6X6 Mesh	Area of the router is 0.26mm^2	–
12	SoCBUS (Wiklund and Liu 2003)	Circuit switching	Distributed	GS, BE	none	—Latency —Network usage —Frequency	Mesochronous	Mesh	–	1–1	Simulation & ASIC (Cycle accurate)	8X8 Mesh	–	–

13	HERMES (Moraes et al. 2004)	Packet switching	XY	BE	Wormhole switching (Header control bits)	—Buffer size vs. number of look up tables (LUT) Buffer size vs. number of gates —Latency —Throughput —Number of clock cycles required to deliver packets —Area	Synchronous	Mesh	–	1-1, One port sending and receiving at the same time	Simulation, FPGA Virtex II	5X5 & 2X2 Mesh	*Vertex II Mapping:* —631 LUTs, -316 Slices *ASIC Mapping:* —2.930 —40,000 transistors —10,000 gates	–
14	BIDI-MIN (Pande et al. 2003) (Bidirectional multilevel interconnect)	Packet switching using VC's	Source	Packet	Wormhole switching	—Throughput vs. number of virtual channels —Silicon area consumed by the switches	Asynchronous	Butterfly fat tree	–	1–1, multicast	Simulation (VHDL)	4X4 Crossbar	*2-input NAND gates Single Channel: 4000 4 virtual channels: 12000*	–
15	OCN (Henriksson, Wiklund, and Liu 2003) (VLSI Implementation of a Switch for On-chip Network)	Circuit switching (round-robin arbitration)	Static	Packet	Header	Functionality	Synchronous	Mesh	–	1–1 (Point-to-point)	Chip manufactured (HP16500 logic analysis)	3X3 Crossbar	Switch area is 0.5 mm²	*Core power consumption*: 0.95 mW (5 MHz) 3.20 mW (50 MHz) *Pad frame power consumption*: 3.55 mW (5MHz) 29.85 mW (50 MHz)

(Continued)

S. No.	Architectures	Comparison Parameters												
		Link Sharing	Routing	QoS	Switching Techniques	Evaluation Parameters	NoC Clocking Mechanism	Topology Supported (2D)	Open Source	Connection Types	Implementation	NoC Size (for experiments)	Area	Power Dissipation/ Energy Consumption
16	SoCIN (Zeferino and Susin 2003)	Packet switching	Deterministic & source based (XY)	Packet	Wormhole switching (header)	—Area vs. FIFO depth (flits) —Area vs. Data width (bits)	Asynchronous	Mesh, Torus	–	–	Synthesize (Altera Quartus II)	–	—Smaller core configuration require 420 logic cells —Larger core configuration require 1754 logic cells	–
17	Star Connected OCN (Lee et al. 2003)	–	–	–	Fast cut-through switching	Top level waveforms of on-chip network	Plesichronous	Star	–	–	–	2X2 Crossbar 16 Master & 16 Slaves	64.8 mm^2	264mW
18	Nostrum (Penolazzi and Jantsch 2006; Millberg et al. 2004)	Packet switching (TDM using loop container (VC's))	Deflective	GS, BE (using deflection routing, it address the congestion control)	Headers	—Average latency —Average bandwidth —Theoretical Hardware study	Synchronous	—Mesh	Yes (on request)	1–1, multicast	Simulation & Synthesis	4X4 Mesh	—A switch with only BE services uses 13695 NAND gates. —With VC capability the gate count increases to 13896	–

19	Nexus (Lines 2004)	Circuit switching	Deterministic	GS	Crossbar	—Frequency —Energy —Latency —Voltage —Area	Globally Asynchronous Locally Synchronous (GALS)	Crossbar	–	1–1, broadcast	—ASIC —Fabricate & Characterized	16X16 Crossbar	—Total area is 4.15 mm² —Crossbar area is 1.75mm² —Pipeline repeater is 0.025mm² —Clock domain converters area is 0.2 mm²	–
20	Xpipes (Bertozzi and Benini 2004)	Packet switching (VC's)	Source-based deterministic (Street sign routing)	Packet	Wormhole switching (Go-back-N)	—Area —Power —Average packet latency vs. bandwidth	Synchronous	Irregular	–	1–1 (Point-to-point)	Simulator & Synthesis (SystemC & XpipesCompiler)	4X4 Crossbar	**—Area (mm²)for the MPEG4 related NoC configuration:** *Mesh:* 1.31 *Custom 1:* 0.86 *Custom 2:* 0.71	**—Power (MW) for the MPEG4 related NoC configuration:** *Mesh:* 114.36 *Custom 1:* 110.66 *Custom 2:* 93.66
21	R²NoC (Samuelsson and Kumar 2004) (Ring Road NoC Architecture)	Packet switching	Distributed and congestion free	Packet	Store and forward	—Traffic distribution stress measurement	Asynchronous	Ring Structure	–	1–1	Simulation (SystemC)	3 Rings with 36 resources	–	–

(Continued)

S. No.	Architectures	Comparison Parameters												
		Link Sharing	Routing	QoS	Switching Techniques	Evaluation Parameters	NoC Clocking Mechanism	Topology Supported (2D)	Open Source	Connection Types	Implementation	NoC Size (for experiments)	Area	Power Dissipation/ Energy Consumption
22	An Architecture and Compiler for aSoC (Liang et al. 2004)	Packet switching using space time scheduling	Stream-based routing	–	Flow control bits	—ASoC vs. hierarchical coreconnect and dynamic network performance —Throughput —Scalability of MPEG2 encoder on ASoC Scalability of image smoothing for 800 × 600 pixel image —Doppler run time for n points	Globally Asynchronous Locally Synchronous (GALS)	Mesh	–	1–1, multicast	—Simulator (timing accurate) —FPGA	9, 16, 25, 36, 49 cores	Tile size ranges from 10000X10000λ² to 30000X 30000 λ²	–
23	μSpider (Evain, Diguet, and Houzet 2004)	Packet switching using VCs (Arbitration= no arbitration, static channel priority & round-robin)	—Deterministic (source) —Dimension order (street sign)	GS & BE	—No back control —Handshake —Credit-based	—Frequency —Area —Traffic class combination cost	Asynchronous	Flexible	–	–	FPGA (Xilink XCV400E)	–	No arbitration: 293 no. of slices Fixed channel priority: 354 no. of slices Round-robin priority: 380 no. of slices	–
24	A 0.13μm NoC (Mondinelli, Borgatti, and KOVACS VAJNA 2004)	Packet switching	Oblivious	–	Store & forward and using header	—Network Throughput —Message latency	Synchronous	Fat tree	–	–	Test chip using HDL description (register transfer level)	64 interfaces(cores)	14 mm²	–

25	Spidergon (Coppola et al. 2004)	Packet switching (VC's)	Deterministic/ shortest path	BE, GS	Wormhole switching (header)	–	Asynchronous	Spidergon	–	–	Simulator (SystemC)	–	–	–
26	RaSoC (Zeferino, Kreutz, and Susin 2004) (Router architecture for SoC)	Packet switching (round-robin arbitration)	Deterministic & source based (XY)	BE	Wormhole switching (also uses handshake protocol)	—Area —Frequency —Cost of buffers —Cost of RaSoc —Cost of bottom-level entities —Number of LCs for FemtoJava	Synchronous	Mesh or torus	–	–	Synthesis (Altera Quartus II 3.0)	–	*-FF based (2 flits)* *8 bit:* 570 LC, 160 Reg *16 bit:* 770 LC, 240 Reg *32 bit:* 1173 LC, 400 Reg *—FF based (4 flits)* *8 bit:* 795 LC, 265 Reg *16 bit:* 1115 LC, 425 Reg *32 bit:* 1830 LC, 745 Reg *—EAB based (2 flits)* *8 bit:* 460 LC, 75 Reg, 100 Mem *16 bit:* 540 LC, 75 Reg, 180 Mem *32 bit:* 700 LC, 75 Reg, 340 Mem *—EAB based (4 flits)* *8 bit:* 486 LC, 90 Reg, 200 Mem *16 bit:* 566 LC, 90 Reg, 360 Mem *32 bit:* 726 LC, 90 Reg, 680 Mem	–

(Continued)

S. No.	Architectures	Comparison Parameters												
		Link Sharing	Routing	QoS	Switching Techniques	Evaluation Parameters	NoC Clocking Mechanism	Topology Supported (2D)	Open Source	Connection Types	Implementation	NoC Size (for experiments)	Area	Power Dissipation/ Energy Consumption
27	QNoC (Bolotin et al. 2004) (Quality of Service NoC)	Packet switching (round-robin arbitration)	Shortest path XY, YX	GS	Wormhole switching (credit-based back pressure)	—ETE delay (uniform traffic) —ETE delay (non-uniform traffic) —Traffic load vs. ETE delay (uniform traffic) —Traffic load vs. ETE delay (non-uniform traffic) —Area —Power	Synchronous	Irregular Mesh	–	–	Simulator (OPNET)	4X4 Mesh	Router area is 10K flip flops	PNoC, Uniform = 1.2 P0
28	On-Chip Network for Low Power Heterogeneous SoC Platform (Lee et al. 2004)	–	–	–	–	–	Globally Asynchronous Locally Synchronous (GALS)	Star	–	–	–	16X16 Crossbar	25mm^2	51mW —31 pj/packet energy consumption
29	NoCGen (Chan and Parameswaran 2004)	Packet switching	Dimension order routing	Packet	Wormhole switching	—Latency —Equivalent Gates	–	Mesh	–	–	Simulator (SystemC)	2X2 4X4 Mesh	–	–

30	Reconfigurable Network on Chip (Ching, Schaumont, and Verbauwhede 2004)	Packet Switching (Round-robin)	–	–	Wormhole switching	–	Synchronous	Torus	Code generator & network source file (open source)	–	GEZEL simulation framework	–	—Area (Slices) 1D Router: 253 2D Router: 674 1X2 1D torus: 531 1X4 1D torus: 1061 2X2 2D torus: 2733	–
31	DyAD (Hu and Marculescu 2004)	Packet switching	—Switches between deterministic & adaptive —XY, odd-even, oe-fixed, DyAD-OE	–	Wormhole switching	—Latency —Throughput —Area	–	Mesh	–	1–1 (Point-to-point)	Simulator & FPGA (Worm_sim, Cycle accurate)	4X4 Mesh 6X6 Mesh 8X8 Mesh (5X5 Cross bar)	—DyAD-OE routers requires 25971 gates. —Odd-even router requires 25,891 gates	–
32	A Routing Switch for On-Chip Interconnection Networks (Chi and Chen 2004)	Round-robin arbitration	Deterministic	–	–	—Latency —Throughput	Synchronous	—Mesh —Torus	–	–	—Simulator (using C++) —TSMC	4X4, 6X6, 8X8 Torus	Area of routing switch is 1.7mmX1.7mm	0.8W
33	Asynchronous On-Chip Network Router with Quality of Service (Feliciian and Furber 2004)	Packet Switching (Virtual Channels)	Dimensional ordered	GS, BE	Wormhole switching with credit-based flow control	—Throughput —Latency —Jitter	Asynchronous	Mesh	–	–	Simulation	5X5 cross bar	–	–
34	Black-Bus (Anjo et al. 2004)	Packet Switching	e-cube routing (any deterministic/ adaptive routing can be applied to black-bus)	–	Store & forward	–	–	Mesh	–	–	Simulator	5X5 Crossbar	–	–

(Continued)

S. No.	Architectures	Comparison Parameters												
		Link Sharing	Routing	QoS	Switching Techniques	Evaluation Parameters	NoC Clocking Mechanism	Topology Supported (2D)	Open Source	Connection Types	Implementation	NoC Size (for experiments)	Area	Power Dissipation/ Energy Consumption
35	SNA (Lee, Lee, and Lee 2004) (SoC Network Architecture)	Packet switching	–	–	–	Cycles vs. IP modules	–	Crossbar, Star Mesh	–	–	Simulation	Cross bar routers	–	–
36	Mango (Bjerregaard and Sparso 2004, 2005a, 2005b) (Message Passing Asynchronous Network on Chip providing Guaranteed services over OCP Interface)	Packet switching	Source routing scheme along with XY	GS, BE, support of GALS	GS: Statistically programmed connection BE: headers and credits	—end-to-end latency —Bandwidth utilization —Throughput —Area —Frequency	Asynchronous (Supports GALS)	Mesh (Not mentioned, concluded from figure)	–	1–1, multicast	Synthesis	5X5 Mesh	0.188mm²	—Zero dynamic power consumption when idle
37	Wolkotte et al. (Wolkotte et al. 2005)	Circuit switching	Distributed	GT, BE	Separate Wire	—Power Estimation —Router Area —Energy consumption —Frequency	Asynchronous	Mesh	–	1–1	Simulation & Synthesis	16X20 Crossbar	Router area is 0.0506 mm²	Power consumption is 3.5 times less as compared to packet switched router
38	ASPIDA (Amde et al. 2005) network on chip	–	–	–	–	—Throughput vs. load —Average latency vs. load	Asynchronous	Custom	–	–	Simulator	–	0.63mm²	–

| 39 | Topology Adaptive NoC (Bartic et al. 2005) | Packet switching (acceptance-dependent round-robin) | Deterministic routing with limited adaptivity | Packet | Virtual cut-through | —arbiter size, LUT and slices vs. number of input ports
—crossbar sizes, LUT and slices vs. number of input ports
—router size, LUT and slices vs. number of router IO's
—network size, LUT and slices vs. number of routers
—network size, LUT and slices vs. network topology
—total network bandwidth vs. network topology | Asynchronous | – | – | – | FPGA (Xilinx Virtex II Pro) | 3X4 Crossbar | Crossbar takes 165 slices (2300 equivalent gates) | – |
| 40 | Spatial Division Multiplexing NoC (Leroy et al. 2005) | Packet switching using spatial division multiplexing (SDM) | Deterministic | GS | – | —Area
—Power
—Frequency | Synchronous | Mesh | – | – | Simulation & Synthesis (Synapsys) | 4X4 Mesh —(pxm)x(pxm) crossbar Where p is total number of ports & m is group of wires | 0.135mm² | 1.79mW |

(*Continued*)

S. No.	Architectures	Comparison Parameters												
		Link Sharing	Routing	QoS	Switching Techniques	Evaluation Parameters	NoC Clocking Mechanism	Topology Supported (2D)	Open Source	Connection Types	Implementation	NoC Size (for experiments)	Area	Power Dissipation/ Energy Consumption
41	Arteris (Arteris 2005)	Packet switching	–	BE	Wormhole switching	—Area —Latency —Frequency —Power	Globally Asynchronous Locally Synchronous (GALS)	Custom	–	1–1 (Point-to-point)	Simulator & ASIC (Electrical)	—72 nodes/IP's (Having 36 masters & 36 slaves) —two 4X5 switches in each cluster for three level of pipelining.	—210K gates for NoC —8K gates for two 4X5 switches in each cluster for three level of pipelining.	–
42	DyNoC (Bobda et al. 2005) (Dynamic NoC)	Packet switching	S-XY (Surrounding XY)	–	–	–	–	Mesh	–	–	FPGA (Virtex II 1000 & 6000)	3X3 4X4 Mesh	—370 slices —Data width of is 8 to 32 bits	–
43	CDMA Router for On-Chip Switched Networks (Kim, Kim, and Sobelman 2005)	Packet Switching (Virtual Channels)	–	–	–	—Throughput —Frequency —Period —Gate count —Area —Throughput	Synchronous	—Star + Star —Star + Mesh	–	–	—ASIC (CX4000 & CX5000) —Simulation (ModelSim)	–	—From 8 bit to 128 bit payload the area is from 39304.2 µm² to 354877 µm² (0.25µm) —From 8 bit to 128 bit payload the area is from 26390 µm² to 2284480 µm²	–
44	Asynchronous NoC Architecture (Beigné et al. 2005)	Packet Switching	Deterministic & odd-even turn model	GS, BE	Wormhole switching	–	Asynchronous (can be implemented as GALS)	Mesh, Tore, Tree	–	–	Simulator (SystemC & Transaction Level Modeling, TLM)	–	—Node has a complexity of 20,000 gates. —Node area is 0.25mm²	–

45	Adaptive Network on Chip (Lee et al. 2005)	–	Adaptive	–	–	–	Mesochronous	–	–	–	–	–	—Transistor count is 409K	–
46	A Low Latency Router (Kim et al. 2005)	Packet switching (Virtual Channels)	Adaptive	Packet	Wormhole switching	—Contention probability vs. offered load —Average latency vs. injection rate —Energy per packet vs. injection rate —Energy delay product per traffic pattern	–	Mesh & Torus	–	–	Simulation (CycleAccurate)	4X4 Crossbar 8X8 Mesh & Torus	–	–
47	Kavaldjiev et al. (Kavaldjiev et al. 2006)	Packet switching using VC's	Source	GS, BE	Headers	—Latency —Throughput —Number of utilized VC per physical channel —Frequency	Synchronous	Mesh	–	1–1	Simulation (Cycle accurate) & Synthesis	6X6 Mesh	Router area is 0.18mm^2	–
48	DSPIN (Panades, Greiner, and Sheibanyrad 2006) (Distributed Scalable Predictable Interconnect Network)	Packet switching using VC's	Deterministic and deadlock X-first (Y-Firs for response)	—GS, BE —Guaranteed Latency	Wormhole switching	–	Mesochronous	Mesh (size or shape of clusters may vary)	–	1–1 (Point-to-point)	Simulation & Synthesis (Cycle accurate systemC) & Synthesize (VHDL)	10X10 Mesh with clusters (one BE initiator, one BE target, on GS target and one optional GS initiator)	Per-router area is 0.082mm^2	–

(Continued)

S. No.	Architectures	Comparison Parameters											
		Link Sharing	Routing	QoS	Switching Techniques	Evaluation Parameters	NoC Clocking Mechanism	Topology Supported (2D)	Open Source	Connection Types	Implementation	NoC Size (for experiments)	Area	Power Dissipation/ Energy Consumption
49	INoC (Neeb and Wehn 2008) (Irregular Network on Chip)	Packet switching	Dijsktra shortest path	Packet	Wormhole switching	—Average traffic vs. flit injection rate —Average latency vs. flit injection rate	–	Chain (Irregular)	–	–	Simulator (Cycle accurate)	5X8	–	–
50	XGFT (Kariniemi and Nurmi 2006) (Extended Generalized Fat Tree)	Packet switching	Adaptive turn back or deterministic	Packet	Header and timers	—Normalized throughput vs. probability of bit error —Normalized throughput vs. probability of faulty channels (%)	Synchronous	Tree	–	–	Simulator & Synthesis	2X4 switch	—XGFT area without FDAR is 7.886 mm² —With FDAR is 9.111 mm²	–
51	High Throughput NoC Architecture (Bouhraoua and Elrabaa 2006)	–	Adaptive (No routing table)	–	Wormhole switching	—Throughput —Latency —Area	Synchronous	Fat tree & modified fat tree (multistage interconnection)	–	–	Simulation (cycle accurate) & Synthesis	32 and 64 clients	Router area is 0.06mm²	–
52	PNoC (Hilton and Nelson 2006) (Programmable NoC)	Circuit switching	Deterministic	GS	–	—Area —Speed —Latency —Frequency —Power	Synchronous (Routers have common clock and nodes have their own clocks)	Subnets (contains routers and nodes)	–	–	FPGA (Xilinx Virtex II)	–	—3685 slices —Speed ranges from 126MHz to 160MHz)	–

53	CTNOC (Sanusi and Wang 2006) (A Central Caching Network on Chip Communication Architecture Design)	Packet switching	XY	–	–	—Latency —Throughput	Synchronous	Mesh	–	–	FPGA (Xilinx Virtex II Pro)	6x6 Mesh	1230 gates	–
54	Low Latency on-Chip Network (Mullins, West, and Moore 2006)	Packet switching	–	Packet	Wormhole switching	—Average latency vs. throughput	Synchronous	Mesh	–	–	Synthesis (UMC's L180 logic process)	4X4 Mesh	–	–
55	GEXPolygon & GEXSpidergon (Zid et al. 2006)	Packet switching	Adaptive (aloha protocol)	Packet	Wormhole switching	–	–	2D Mesh, Tree & octagon	–	–	Simulator (ModelSim)	8X(2X2)	–	–
56	Low-Power Network on Chip (Lee, Lee, and Yoo 2006)	Packet switching (intercluster) Circuit switching (with in cluster)	Round-robin	Packet	–	—Power consumption —Access count vs. number of transitions between successive data words —Energy consumption	Mesochronous	Hierarchical Star	–	–	–	8X8 crossbar —8 Master —8 Slaves	5X5 mm^2	Chip consumes 160 mW and dissipates 51 mW
57	CoNoChi (Pionteck, Koch, and Albrecht 2006)	Packet switching	Deterministic	–	Virtual cut-through	–	–	Mesh	–	–	FPGA (Virtex II Pro 100)	–	—410 Slices —Data width of is 8, 16 & 32 bits —Latency is 5 clock cycles	–

(*Continued*)

S. No.	Architectures	Comparison Parameters												
		Link Sharing	Routing	QoS	Switching Techniques	Evaluation Parameters	NoC Clocking Mechanism	Topology Supported (2D)	Open Source	Connection Types	Implementation	NoC Size (for experiments)	Area	Power Dissipation/Energy Consumption
58	Cross Road Interconnection Architecture (Chang, Shen, and Chen 2006)	Circuit switching	Deterministic	GS	–	—Throughput —Power —Latency —Characteristics of localization —Area	Asynchronous	Irregular	–	–	Simulator & Synthesis (Modelsim)	–	56.26×10^3 μm^2	260.6 µW
59	HIBI (Salminen et al. 2006) (Heterogeneous IP Block Interconnection)	Circuit & packet switching using Time Division Multiple Access (TDMA, round-robin)	Deterministic	GS	Wormhole switching	—Area —Frequency —Clock cycle vs. transfer size —Area vs. transfer area —Cost vs. transfer size —Latency —Throughput	–	Irregular	–	multicast	Simulator & FPGA (VHDL & SystemC)	–	5400 gates	–
60	On-Chip Multimedia Applications (Lee et al. 2006)	–	–	–	Wormhole switching	—Area —Power —Energy	Synchronous	–	–	–	FPGA (Virtex II 4000)	–	Area is 12,929 slices	Overall power consumption of all the interface, link & router is 1288 mW
61	Dynamic Reconfigurable NoC for Adaptive Reconfigurable MPSoC (Ahmad, Erdogan, and Khawam 2006)	Circuit switching along with packet switching	Adaptive	–	Wormhole switching	—Throughput percentage vs. network load percentage —Time percentage vs. network load percentage —Throughput percentage vs. network type	–	Torus	–	–	Simulator (NS-2)	4X4 Torus (16 nodes)	–	–

62	Communication Architecture Optimization (Ogras et al. 2006)	Packet Switching	–	–	Wormhole switching	—Percentage of links added vs. network size —Total number of links vs. network size —Network diameter vs. network size —Cost factor vs. network size —Utilization vs. router id —Average packet latency vs. packet injection rate (packets/cycle) —Power —Energy —Execution time	–	Mesh	–	–	FPGA (Xilinx Virtex II)	4X4 Mesh —Degree of standard mesh network is 4	Energy consumption is 17.14µJ without long range links and 11.72 µJ with long range links	—Power consumption without long range links is 334.7mW —Power consumption with long range links is 343.2mW
63	ProtoNoC (Castells-Rufas, Joven, and Carrabina 2006)	Ephemeral Circuit Switching (Combination of circuit and packet switching)	XY	GS	Combination of store and forward & virtual cut-through & wormhole	—Bandwidth vs. number of hops —Area —Frequency	Globally asynchronous locally synchronous (GALS)	Mesh	–	–	Simulator & Synthesis (Quartus II Altera S30)	4X4 Mesh	—Logic cells for Synchronous is 5995 —Logic cells for Asynchronous is 5764	–
64	TTNoC (Schoeberl 2007) (Time Triggered Network on Chip)	Pulsed	Contention free and distributed	–	Header	—Area (Logic cells) —Frequency —Memory	Globally Asynchronous Locally Synchronous (GALS)	Ring	–	1–1, broadcast	FPGA (Altera Cyclone II)	–	NI needs 480 logic cells	–
65	NocMaker (Castells-Rufas et al. 2009)	Packet switching & Circuit switching	–	Packet	–	—Throughput —Mean latency —Area —Power	Synchronous	Mesh	Yes	1–1 (Point-to-point)	Simulation & FPGA (JHDL)	3X3 Mesh	–	–
66	TILEPro64 (Bell et al. 2008)	Packet switching	XY	Packet	Wormhole switching with credit-based control	–	Synchronous	Mesh	–	–	Fabrication (Triple-V_k CMOS process)	8X8 Mesh (5X5 Crossbar)	–	–

(Continued)

| S. No. | Architectures | Comparison Parameters | | | | | | | | | | | | | |
		Link Sharing	Routing	QoS	Switching Techniques	Evaluation Parameters	NoC Clocking Mechanism	Topology Supported (2D)	Open Source	Connection Types	Implementation	NoC Size (for experiments)	Area	Power Dissipation/ Energy Consumption
67	UT TRIPS (Gratz et al. 2007)	Packet switching using VC's	YX	Packet	Wormhole switching with credit-based control	—Offered bandwidth —Latency (cycles) —Instructions per cycle	Synchronous	Mesh	–	1–1 (Point-to-point)	Simulator & ASIC (TRIPS processor core & low level)	4X10 Mesh (6X6 Crossbar)	Router area is 1.10 mm^2	–
68	Intel TeraFLOPS (Vangal et al. 2008)	Packet switching using VC's	Source directed	Packet	Wormhole switching with on/off buffer management	—Power —Area —Frequency vs. VCC —Power vs. Number of active ports	Mesochronous	Mesh	–	1–1 (Point-to-point)	—ASIC —Circuit research lab —Application research lab —Software solution group —Logic technology development —Arizona technology development —Mask design team	8X10 Mesh (5X5 Crossbar)	—Total area of 275mm^2 —100M transistors —Router area is 0.34 mm^2 —1 TFLOPs of performance	15.6W to 230W
69	Ambric (Butts 2007)	Circuit switching	–	–	–	–	Synchronous	Mesh	–	–	ASIC	42 CR-RU bricks	—180 million transistors —1.03 trillion operations per second	12.6 MIPS/mW
70	STNoC (Palermo et al. 2007)	Packet switching	–	–	–	–	–	Spidergon	–	–	Simulator (OMNET++)	–	–	–
71	EIB on-Chip Network (Ainsworth and Pinkston 2007)	Circuit switching	–	–	Wormhole switching	—Packet latency	–	–	–	–	–	–	–	–

72	SCC (Hoffman et al. 2007) (Scalable Communication Core)	Packet switching	XY, YX, Odd-even	Packet	Wormhole switching & header	—Latency —Throughput —Area —Power	Asynchronous	Mesh	–	–	Simulator	3-ary 2-cube (5x5 Crossbar)	0.8mm²	18mW
73	MoCRes (Janarthanan, Swaminathan, and Tomko 2007)	Packet switching	XY	Packet	Virtual cut-through	—Area —Latency	Multi-Clock	Mesh	–	–	FPGA (Virtex-4)	3X3 Mesh	282 Slices (558 LUT's, 289 FF's)	–
74	Generalized de Bruijn Graph NoC (Hosseinabady et al. 2007)	Packet switching	Source	Packet	Wormhole switching (Header)	—Average shortest paths —Latency —Energy consumption —Area —Power —Latency(delay)	–	Generalized de Bruijn graph	–	–	Simulation (ModelSim) Synthesis (Synopsys)	–	—Area(μm², 0.18μm, VDD=1.8V) Normal switch: 0.406 Reliable Switch: 0.588 —Area(μm², 0.13μm, VDD=1.2V) Normal switch: 0.185 Reliable Switch: 0.272 —Area(μm², 0.09μm, VDD=1V) Normal switch: 0.100 Reliable Switch: 0.147	—Power (0.18μm, VDD=1.8V) Normal switch: Dynamic(mW)= 25.8 Static (μW)=1.67 Reliable Switch: Dynamic(mW)= 38.5 Static (μW)=2.43 —Power(0.13μm, VDD=1.2V) Normal switch: Dynamic(mW)= 3.87 Static (μW)=50.52 Reliable Switch: Dynamic(mW)= 5.59 Static (μW)=75.87 —Power(0.09μm, VDD=1V) Normal switch: Dynamic(mW)= 2.54 Static (μW)=127.69 Reliable Switch: Dynamic(mW)= 3.98 Static (μW)=187.07

(Continued)

S. No.	Architectures	Comparison Parameters												
		Link Sharing	Routing	QoS	Switching Techniques	Evaluation Parameters	NoC Clocking Mechanism	Topology Supported (2D)	Open Source	Connection Types	Implementation	NoC Size (for experiments)	Area	Power Dissipation/ Energy Consumption
75	Polaris (Soteriou et al. 2006)	Packet switching	Adaptive & deterministic	Packet	Wormhole switching & virtual cut-through	—Pipe length (cycles) —Hop length (cycles) —Latency —Network power —Energy/flit —Energy Delay/ flit —Power/Area	Synchronous	Ring (1D), Binary fat tree, 2D and 3D mesh, torus	–	–	Synthesis (ModelSim SE 5.6)	64 nodes (NXN Mesh)	—310 mm² —Area is 1e+7µm² to 7.80e+6 µm² —3.22 10^{-11} to 44.3310^{-11}J Energy/flit —1.87 10^{-19} to 50.74 10^{-19} JS	Network power is 4.98W to 68.58W
76	CSRA-NoC (Braun et al. 2007) (Circuit Switched Run Time Adaptive NoC)	Circuit switching	–	–	–	—Area —Frequency	–	Mesh	–	–	FPGA (Virtex II)	–	—Switch uses 100 slices —Data width of is 8 bit —Setup	–
77	artNoC (Schuck, Lamparth, and Becker 2007) (Adaptive Real-Time Network on Chip)	Packet Switching (VC's)	XY & west first	BE, GS	Wormhole switching	—Area	–	Mesh	–	Multicast	FPGA (Virtex II Pro)	2X2	—0.519mm²	–
78	CDMA NoC (Wang, Ahonen, and Nurmi 2007)	Packet switching	CDMA	Guaranteed communication	Wormhole switching	—Area —Power —Energy	Globally asynchronous locally synchronous (GALS)	Ring	–	Multicast	Synthesize	6 Nodes	—Area (Kilo gates) from 113.145 (1 bit) to 272.806 (32 bits) —Spreading codes are 8 bit walsh codes	—Dynamic Power (mW) from 19.340 (1 bit) to 7.332 (32 bit) —Energy (pj) from 12.5168 (1 bit) to 4.0873 (32 bit)

79	A Reconfigurable Baseband Platform Based on Asynchronous NoC (Lattard et al. 2008)	Packet Switching (one VC for minimum latency while another is used for high throughput with no guarantee of latency) —Round-robin arbitration	Odd-even turn model	–		Wormhole switching with credit-based flow	—Area —Power	Asynchronous (24 on-chip clocks)	Mesh	–	–	–	Mesh (20 nodes)	—Chip area is 79.5mm² —More than 3M gates and 3.5 Mb of embedded RAM —NoC area is 15% of the globe area —Core area is 72.71mm²	—Chip power consumption 640mW Transmitter mode and 760mW receiver mode —NoC power consumption is 6% of the global consumption —Dynamic frequency scaling
80	TTNoC (Paukovits and Kopetz 2008) (Time Triggered Network on Chip)	Pulsed	Source along with contention free	GS		Two control wires (valid and header)	—Introduction of period & phases —end-to-end latency —Topology —Switching —Energy & die area —Level of abstraction (No graphs or quantitative results are provided)	Synchronous	Ring	–	1–1, multicast	FPGA (Altera Cyclone II)	2X3 Mesh	Total logic elements is 444	–

(*Continued*)

S. No.	Architectures	Comparison Parameters												
		Link Sharing	Routing	QoS	Switching Techniques	Evaluation Parameters	NoC Clocking Mechanism	Topology Supported (2D)	Open Source	Connection Types	Implementation	NoC Size (for experiments)	Area	Power Dissipation/ Energy Consumption
81	EVC (Kumar et al. 2008) (Express Virtual Channels)	Packet switching	XY	Packet	EVC	—Normalized latency —Normalized energy	Asynchronous	Mesh	–	1–1, multicast	Simulator (Cycle accurate)	7X7 Mesh having 2 hop EVC	–	–
82	ReNoC (Stensgaard and Sparsø 2008) Reconfigurable Network on Chip	Packet switching plus circuit switching	–	Packet	Header	—Area —Power —Latency	Synchronous	Any	–	–	Simulation (Spice) Synthesize (FPGA)	Mesh	—Area of Router (mm^2) Static NoC 0.53 ReNoC mesh: 0.58 ReNoC specific: 0.58	—Power of Router (mW) Static NoC: 4.56 ReNoC mesh: 4.69 ReNoC specific: 2.02 —Energy consumption: 21pj/packet
83	MoCSYS (Janarthanan and Tomko 2008) (Multi-Clock SYS)	Packet switching plus circuit switching	XY	GT	Virtual cut-through	—Bandwidth —Area	Multi-Clock	Mesh	–	–	FPGA (XC4VLX100)	–	–	–
84	Aelite (Hansson, Subburaman, and Goossens 2009)	Packet switching (TDM)	Source along with contention free	GS	Headers	—Frequency —Area	Globally Synchronous with mesochronous or asynchronous links	—Mesh	No	1–1, channel tree	Synthesis	4X3 Mesh	Arity-5 router requires $0.032mm^2$ area	–

85	HT-OCTAGON (El-Moursy, Korzec, and Ismail 2009) (High Throughput OCTAGON)	Packet switching (VC's)	Source	Packet		Wormhole switching (four separate control lines)	—Frequency —Throughput —Area	Asynchronous	Irregular	–	1–1	ASIC	8X8 arbiter, 8 nodes, 12 bidirectional buses	—110000 number of transistors are required for VC=4 —Area is decreased by 18% as compared to OCTAGON —Total wires needed is 7057.92mm	Switch consumes 22072.3mW
86	XHiNoC (Samman, Hollstein, and Glesner 2009) (Extendable Hierarchical NoC)	Packet switching (round-robin arbitration)	Static (XY) and adaptive (west first, east last, negative first, odd-even)	Packet		Wormhole switching	—Area —Power —Frequency	Synchronous	Mesh	–	1–1, multicast	FPGA (Virtex II pro, xc2vp30)	4X4	—Static routing (FIFO=2 register): 0.0767mm² (FIFO=4 register): 0.1018mm² —Adaptive routing (FIFO=4 register): 0.106mm²	—*Net switching (mW)* XY: 7.49 WF: 7.94 NF: 8.04 —*Cell internal power (mW)* XY: 63.98 WF: 60.87 NF: 66.43 —*Cell leakage power (µW)* XY: 0.89 WF: 0.99 NF: 0.84
87	Network on Chip in a Three Dimensional (Feero and Pande 2008)	Packet switching	–	Packet		Wormhole switching	—Energy —Throughput —Latency	Synchronous	Butterfly fat tree	–	–	Simulator (Cycle accurate)	64 nodes & 28 switches	–	–

(Continued)

S. No.	Architectures	Comparison Parameters												
		Link Sharing	Routing	QoS	Switching Techniques	Evaluation Parameters	NoC Clocking Mechanism	Topology Supported (2D)	Open Source	Connection Types	Implementation	NoC Size (for experiments)	Area	Power Dissipation/ Energy Consumption
88	BiNoC (Lan et al. 2011) (Bidirectional NoC)	Packet switching (VC's)	XY or Odd-even turn model	Packet	Wormhole switching	—Frequency —Latency vs. flit injection rate —Direct switching probability vs. flit injection rate —Bandwidth utilization vs. flit rate —Latency vs. depth of each buffer —Area —Power	Synchronous	Mesh	–	–	Simulator (Cycle accurate)	—8X8 Mesh —6X6 Mesh —5X5 Mesh —4X4 Mesh (-5X5 Crossbar —10X10 Crossbar)	*BiNoC (No VC)* 48666 gates *BiNoC (4VC)* 58681 gates	*BiNoC (No VC)* 53.21mW *BiNoC (4VC)* 44.4mW
89	ALPIN (Beigné et al. 2009) Asynchronous Low Power Innovative NoC	Packet switching	–	GS, BE	–	—Throughput —Latency —Leakage	Globally Asynchronous Locally Synchronous (GALS)	Mesh	–	–	—STMicro—electronics test board —FPGA	2X3 Mesh	11.56mm²	*High@1.2V (mW)* FHT: 20.12 MEM: 32.63 OFDM: 81.07 *Low@0.8V (mW)* FHT: 2.66 MEM: 4.31 OFDM: 10.07 *Hopping@ 50% (mW)* FHT: 12.3 MEM: 19.4 OFDM: 46.4 *IDLE@1.2V (µW)* FHT: 682 MEM: 883 OFDM: 632

90	RAMPSoC (Göhringer et al. 2010) (Run-time Adaptive Multiprocessor System on Chip)	Heterogeneous communication (packet switching plus circuit switching)	–	Packet	–	–	Different clock domains	Heterogeneous (star-wheels, combination of Spidergon and star)	–	1–1 (Point-to-point)	Simulator & FPGA (PLI, C++ and HDL combination)	9 Agent Micro Network	IP core size of 20K to 200K gates	–
91	DRNoC (Krasteva, De la Torre, and Riesgo 2010) (Dynamic Reconfigurable NoC)	Packet switching	XY & table base routing	Packet	Wormhole switching	—Area —Speed —Frequency —Area vs. number of router ports	Asynchronous	—XMesh (Reconfigurable router, NI and PE)	–	1–1 (Point-to-point), Multicast (point to multipoint)	FPGA (Virtex II)	—4X2 NoC —6X6 Crossbar	—Speed (391 slices for 180 MHz) —Area (385 slices for 129 MHz) —BRAMs (CLB's is 5)	–
92	PMCNOC (Wang et al. 2010)	Packet switching	XY	–	–	—Latency —Throughput —Area —Frequency	Synchronous	Mesh	–	–	FPGA (Xilinx Virtex II Pro)	6x6 Mesh	1230 gates	–
93	Dynamic Reconfigurable Network on Chip (Wu, Tang, and Hsu 2011)	Packet switching	North-last-weave	Packet	Wormhole switching	—Latency vs. injection rate (flit/cycle/node)	Asynchronous	Mesh	–	1–1, multicast	Simulator (Cycle accurate simulator)	8X8 Mesh	–	–
94	Ramos et al. (Peña-Ramos and Parra-Michel 2011)	Circuit switching	Static	Packet	Credit mechanism	—Setup time takes 400 clock cycles	Asynchronous	–	–	1–1	—Simulation (QuestaSim) —Virtex 5 FPGA	8X8 cross bar	–	–

(Continued)

S. No.	Architectures	Comparison Parameters												
		Link Sharing	Routing	QoS	Switching Techniques	Evaluation Parameters	NoC Clocking Mechanism	Topology Supported (2D)	Open Source	Connection Types	Implementation	NoC Size (for experiments)	Area	Power Dissipation/ Energy Consumption
95	ORNoC (Le Beux et al. 2011) (Optical Ring Network on Chip)	Wavelength Division Multiplexing (WDM)	Wavelength routing with contention free	–	–	–	–	–	–	Intercluster/ interlayer communication	–	2D and 3D architectures 2D: 3X3 clusters 3D: 2X2 clusters per layer —1296 nodes —102 waveguides and 64 wavelength per waveguide	–	–
96	Kilo-NoC (Grot et al. 2011) (Thousand Connected NoC)	Packet switching (VC's)	Look-ahead	Packet	Virtual cut-through	—Area —Average packet latency —Fairness & throughput —Average packet slow down —Power efficiency	–	Multidrop Express Channel (MECS)	–	1–1 (Point-to-point), Multicast (point to multipoint)	Simulation (Orion, CACTI)	1024 terminals with 256 concentrated nodes (64 shared resources)	256mm^2	23W (when 10% load)

97	WiNoC (Ganguly et al. 2010)	Packet switching	Shortest path along with distributed routing	Packet	Wormhole switching	—Latency —Throughput —Energy dissipation	Synchronous	Small-world (subnets having cores)	–	Multicast	Simulator (Cycle accurate simulator)	128, 256 and 512 cores on a die size of 20mm x 20mm	Total area of WiNoC switches and hubs ranges from approx. 28mm2 to approx. 80mm2 for different number of subnets and cores	WiNoC packet energy dissipation ranges from 22.57nJ to 37.48nJ for different system size
98	dAElite (Stefan, Molnos, and Goossens 2012) (distributed AElite)	Packet switching (TDM)	Contention free and distributed	GS	TDM with separate wires (credit-based flow control at NI)	—Frequency —Area (with TDM wheel size) —Synopsys TSMC 65nm —Cost of dAElite as compare to other NoC architectures	Synchronous	Any	No	1–1, multicast	Synthesis (FPGA)	2X2 Mesh	0.17mm² (for 32 slots)	–
99	BMNoC (Lee et al. 2012) (Busmesh Network on Chip)	Packet switching	Wormhole	Packet	Wormhole switching using credit mechanism	—Latency vs. injection rate (flit/cycle/ node)	Asynchronous	Local buses and Mesh	–	1–1	Simulator (NS-2)	2x2 Mesh with clusters	–	–

(Continued)

S. No.	Architectures	Comparison Parameters												
		Link Sharing	Routing	QoS	Switching Techniques	Evaluation Parameters	NoC Clocking Mechanism	Topology Supported (2D)	Open Source	Connection Types	Implementation	NoC Size (for experiments)	Area	Power Dissipation/ Energy Consumption
100	Custom Network on Chip Architecture (Mishra, Nidhi, and Kishore 2012)	Packet switching	–	Packet	Virtual cut-through (header)	–Resource utilization	Asynchronous	Tree	–	1–1	FPGA (Xilinx Virtex-5)	One main processor along with four mapping processors	Look up table: NIA:231 NIB:189 MOD1:260 MOD2:231 SLICES: NIA:228 NIB:192 MOD1:127 MOD2:104 Resource utilization in terms of 3 and 5 processors are also mentioned	–
101	DANoC (Shu et al. 2012) (A dynamic adaptive Network on Chip Architecture)	Packet switching	XY	Packet	Wormhole switching	–Latency vs. injection rate –Throughput vs. injection rate	Asynchronous	Simple & regular (Mesh)	–	1–1	Simulator (Cycle accurate simulator based on NOXIM)	4X4 8X8	–	–
102	WaveSync (Yang et al. 2014)	Packet switching	Right turn	Packet	Wormhole switching using credit mechanism	–Latency –Throughput –Area –Power	Globally Asynchronous Locally Synchronous (GALS)	Mesh	–	–	Synthesize (TSMC)	7X7 Mesh	–For 600MHz: 0.124mm² –For 2000MHz 0.131mm²	–For 600MHz: 34.87mW –For 2000MHz 126.95mW

103	AdNoC (Al Faruque, Ebi, and Henkel 2010) (Adaptive Network on chip)	Packet switching (VC's)	Run-time route allocation (weighted XY)	Packet	Wormhole switching	—Unsuccessful connection vs. time interval (ms) —Maximum link capacity vs. different mapping instances —Successful transactions & corresponding buffer requirement	Synchronous	Mesh	—	—	Synthesize (Xilinx Virtex2 XC2V-6000 FPGA)	4X4 Mesh	—*wXY-routing*: 122 slices —*On-demand VCB*: 129 slices —*Monitoring component*: 46 slices —*Total area*: 297 slices	725mW
104	Aurora (Qouneh et al. 2012)	Photonic circuit switched along with electrical packet switching	Shortest distance (SD), shortest path (SP) first and temperature first (TF)	—	Virtual channel flow control	—Latency —BER and MER —Power Consumption	Synchronous	Folded torus hybrid	—	—	Simulation (Simics & GARNET)	5X6	Total chip area is 20mm X 24mm	—Power consumed by each modulator is 200 fj/bit —In plane Poly-Si energy consumed is 100 fj/bit
105	Mesh-based NoC (Choudhary and Qureshi 2012)	Packet switching	Cross routing, Dyad, fully adaptive, negative first, north last, odd-even, west first, XY	Packet	Wormhole switching	—Latency —Throughput —Energy	Asynchronous	Mesh & Partial interconnected mesh	—	—	Simulator (Noxim & Nirgam)	4X4 & 8X8 Mesh	—	—

(Continued)

S. No.	Architectures	Comparison Parameters												
		Link Sharing	Routing	QoS	Switching Techniques	Evaluation Parameters	NoC Clocking Mechanism	Topology Supported (2D)	Open Source	Connection Types	Implementation	NoC Size (for experiments)	Area	Power Dissipation/ Energy Consumption
106	RecMIN (Logvinenko, Gremzow, and Tutsch 2013) (Reconfigurable Multi-Interconnection Network)	–	Backpressure	Packet	–	—Latency —Throughput	Synchronous	MIN (8X8 inputs/ outputs built by 2X2 routers or with 4X4 router with four simple wire connections) —NoC having 2^n sizes (n=1,2,...)	–	1–1 (Point-to-point)	Simulation (RecSim)	NoC having 16 inputs and outputs	–	–
107	SWIFT (Postman et al. 2012) (Swing reduced Interconnect for a Token-based NoC)	Packet switching (VC's)	Minimal routing with west first turn rule	Packet	Token	—Average packet latency —Power	Synchronous	Mesh	–	–	Simulation & Synthesize (2X2 test chip, RTL)	8X8 (5X5 Crossbar)	Router area is 0.48 mm²	–
108	SpiNNaker (Furber et al. 2012)	Packet switching	Multicast	BE	Header (control byte)	–	Asynchronous	Mesh	–	Multicast	Simulation	—Upto 65536 nodes —Two NoC routers	–	90KW
109	H-NoC (Carrillo et al. 2012)	Packet switching	Region base	BE	Header	-FiringFrequency—	Asynchronous	—Mesh —Star	–	Unicast, Multicast, broadcast	Simulation & FPGA	—One cluster node connected with four tile nodes. The four nodes are connected with 10 nodes. —40 nodes at the bottom of the hierarchy	0.587 mm2	13.16 mW

#	Name													
110	BLOCON (Kao and Chao 2012) (Buffer less photonic CIOs Network)	Packet switching	–	–	Wormhole switching using credit mechanism	—Laser Power Calculation —Cost Evaluation —Latency vs. injection rate	–	Mesh	–	Many to Many	Simulation	64X64	—Area of 0.35mm² —Ring number=39K	-Thermal Tuning Power=0.39W —Energy consumption per link is 36.8 pJ
111	HELIX (Bahirat and Pasricha 2014) (Hydro nano photonic-electric network on chip)	–	XY along with modified adaptive photonic concentration region routing	–	–	-Power (mW) —Throughput (flit/cycle) —Latency (ns)	–	Irregular	–	1–1 (Point-to-point)	Simulation (Cycle accurate derived from NOXIM)	–	400mm²	–
112	QORE (DiTomaso, Kodi, and Louri 2014)	Packet switching	XY with deadlock avoidance technique	–	–	-Dynamic power —Leakage power —Link Failures —Channel utilization —Discharge time of channel buffer —Area overhead —Power overhead	(2 GHz)	-Mesh	–	–		4X4	793,442 µm2	413.2 mW
113	DMesh (Wang and Bagherzadeh 2014)	Packet switching	XY (fixed priority)	GT, BE	Wormhole switching	-Average latency vs. traffic load —Throughput		Mesh	–	–	Synthesis (Synopsys Design Compiler)	16X16 32X32	56,583 µm²	Dynamic power: 33.68 mW Leakage power: 265.02 µW
114	STNR (Poluri and Louri 2014)	Packet switching	XY	–	Header	-Average latency	–	Mesh	–	–	–	-For benchmark traffic: 4X4 mesh —For synthetic traffic: 8X8 mesh	Incurs an area overhead of 7% with respect to the baseline router	Incurs power overhead of 13% with respect to the baseline router

(Continued)

S. No.	Architectures	Comparison Parameters												
		Link Sharing	Routing	QoS	Switching Techniques	Evaluation Parameters	NoC Clocking Mechanism	Topology Supported (2D)	Open Source	Connection Types	Implementation	NoC Size (for experiments)	Area	Power Dissipation/ Energy Consumption
115	inoc (Heisswolf et al. 2015)	Packet switching	XY plus odd-even	GT, BE	Wormhole switching	-Detection rate —Recognition time —Throughput —Average latency	–	Mesh	–	–	Synthesis	4X4, 8X8	For single router with 32 bit link (µm2): VC4: 1.5 VC6: 1.7 VC8: 2	–
116	TagNoC (Yaghini, Eghbal, and Bagherzadeh 2015)	Packet switching	Source and distributed	BE	Wormhole switching	-Packet latency vs. offered load —Power consumption —Area	Network: Synchronous PEs: asynchronous (1GHz clock)	Mesh, Torus	–	–	Synthesis (Synopsys Design Compiler)	4X4	Router area si 138 µm²	-Router power cosumption: 0.009 mW —NI power over head is 21.37 µW
117	Optical Multilevel Signaling (OMLS) NoC (Kao and Louri 2016)	Packet switching	–	–	–	-Bandwidth -Power -Area	10 GHz	Clos	–	–	Synthesis	64 tiles	–	–
118	Mesh Star Mesh (MSM) Architecture (Das and Ghosal 2017)	Packet switching	XY	BE	Wormhole switching	-Latency —Area	–	Mesh	–	–	Simulation (Noxim)	4x4 6x6 8x8	Each PE area is 1.56 mm²	–
119	Phosphorus Architecture (Mediouni and Hasnaoui 2017)	Packet switching (round-robin arbitration)	Dijkstra's algorithm	–	Wormhole switching	-Area —Power consumption	500 MHz	3D Mesh	–	–	Simulation (BookSim2)	4x4x3 (48 nodes)	0.863649 mm2	8.53 W
120	DAPPER NoC Architecture (Raparti and Pasricha 2018)	Packet switching	XY	–	Wormhole switching	-Latency —Energy consumption	–	Mesh	–	–	Simulation (Noxim)	4x4 (16 cores)	–	–

121	ShuttleNoC Architecture (Lu et al. 2018)	Packet switching	–	–	-Packet network latency —Energy consumption —Power consumption	0.8 GHz	Mesh Torus	–	–	Simulation (Cycle accurate)	4x4 8x8 16x16	–	–
122	TAONoC Architecture (Yang et al. 2018)	Optical Signaling Communication	–	–	—number of MRRS vs. number of tiles —number of waveguides vs. number of tiles —maximum insertion loss vs. number of tiles —average insertion loss vs. number of tiles	2 GHz	Torus	–	–	Simulation (OPNET)	4x4 8x8 16x16	–	–

Survey of noc architectures: Study 2

SNo.	Architectures	Comparison Parameters									
		Year	Flit Size	Buffering	PE-Router Interface	Estimated Peak Performance	Fault Tolerance	NoC Frequency/ Technology	Switch/ Router Ports	Bus/ Channel width	Latency (per router/switch/ NoC)
1	SPIN (Guerrier and Greiner 2008) (Scalable Programmable Interconnect Network)	2000	32 bits (plus 4 control bits)	Input queue plus 2 shared output queue (1 virtual channel)	VCI	2Gbits/s (per switch)	–	200 MHz, 0.25 μm	8	Channel width is 36 bits (32 data bits plus 4 tag bits used for packet framing, parity and error signaling)	It has a latency of 30 cycles and saturates at the offered load of 28%
2	aSOC (Liang, Swaminathan, and Tessier 2000) (Adaptive System on Chip)	2000	32 bit	Input & output buffers	–	–	–	333 MHz, 1.2 micron	5	–	-Write/Read operation takes 5 clock cycles —Critical path in SPICE simulation is 8.5ns
3	Dally et al. (Dally and Towles 2001)	2001	-Number of flits in the packet are 16 —Flit size is 16 bit	Input and output buffers (size is 10K bits)	–	4Gb/s	Transient faults (error in bits) detection mechanism	2GHz or 200MHz, 0.1μm	5	16 bit link	–
4	MicroNetwork (Wingard 2001)	2001	128 bit data word	Input and output FIFO	OCP	–	–	250 MHz, 0.18μm	–	–	–

5	OCTAGON (Karim, Nguyen, and Dey 2002)	2002	Variable data (plus 3 control bits)	Input buffers (less than 50 packets)	–	-40 Gbits/s —Saturation point of 0.90 flit/cycle/node for VC=8	–	500MHz	–	Width of the link is 32 bits	–
6	CHAIN (Bainbridge and Furber 2002)	2002	16 bit header plus 8,16,32 bit data	Pipeline latches	OCP	-700Mbits/s per link for 0.35-micron —1Gbits/s per link for 0.18-micron CMOS	–	Over 100 MHz, 0.18 micron 0.35 micron	3 input ports and 4 output ports	–	Latency is 3ns per symbol on 0.35 micron —Total time of 26ns is required for one route setup and read data.
7	PROTEO (Siguenza-Tortosa and Nurmi 2002)	2002	Variable data and control sizes	Input and output buffers	-PVCI —BVCI —AVCI	–	Fault Tolerant (No detail provided, Trying to include)	–	3	–	–
8	CLICHÉ (Kumar et al. 2002) (Chip Level Integration of Communication Heterogeneous Elements)	2002	Packet size is 256 bits	Input queue (size is 4 packets)	–	200Mbits/sec	–	60nm	5	Data bus width is 256 bits	–
9	RAW (Taylor et al. 2002)	2002	Data bus is 32 bit plus 5 control line	Output FIFO	–	-25Gbps of input & output bandwidth —On-chip memory bandwidth of 57Gbps	–	225 MHz, 0.15-micron 180nm	–	32 bit data bus	Router takes one clock cycle per node

(Continued)

SNo.	Architectures	Comparison Parameters									
		Year	Flit Size	Buffering	PE-Router Interface	Estimated Peak Performance	Fault Tolerance	NoC Frequency/Technology	Switch/Router Ports	Bus/Channel width	Latency (per router/switch/NoC)
10	Eclipse (Forsell 2002)	2002	–	Output queue	–	–	–	–	5	–	Superswitch takes 1 clock cycle for switching
11	Æthereal (Rijpkema et al. 2003)	2003	Packet size is 8 flits (data path is 34 bits including 2 control bits)	Input queue GS queue size is 1 flit, BE queue size is 8 flits —256 slots	–	*-GS-BE distributed programming router:* 2Gbytes/s *—GS-BE centralized programming router:* 4Gbytes/s —Aggregated throughput of 80Gbits/s	–	500MHz, 0.13 μm	5	–	–
12	SoCBUS (Wiklund and Liu 2003)	2003	–	–	–	64GByte/s	–	SoCBUS clock of 1.2 GHz and IP block clock frequency of 300 MHz, 0.18 micron	5	–	-Four bus cycles per switch for request signal —Acknowledgment signal latency is 1 cycle per switch —Data transfer latency is 0.25 cycles

#	Name	Year										
13	HERMES (Moraes et al. 2004)	2003	8 data bits (plus 2 control bits) Parametrizable	Input queue	OCP	500 Mbits/s (per switch)	–	25MHz, 0.35μm	5	Link width is 8 bits	-Header flit takes 10 clock cycles per switch —Data flits take 2 clock cycles per switch
14	BIDI-MIN (Pande et al. 2003) (Bidirectional multilevel interconnect)	2003	42 bits	Input & output queues	–	Switches can received and send data at GHz range	–	0.18μm	6	–	–
15	OCN (Henriksson, Wiklund, and Liu 2003) (VLSI Implementation of a Switch for On-chip Network)	2003	4 data bits plus 2 control bits	–	–	–	–	50MHz	3	–	–
16	SoCIN (Zeferino and Susin 2003)	2003	Flit size is n+2 (where n is 8, 16, 32 bits)	Input buffers (1,2,3 or more words)	OCP, VCI	–	–	73 MHz to 49 MHz	5	Channel width is 8, 16, 32 bits	–

(Continued)

SNo.	Architectures	Comparison Parameters									
		Year	Flit Size	Buffering	PE-Router Interface	Estimated Peak Performance	Fault Tolerance	NoC Frequency/ Technology	Switch/ Router Ports	Bus/ Channel width	Latency (per router/switch/ NoC)
17	Star Connected OCN (Lee et al. 2003)	2003	Packet size is 84 bits	–	–	Peak throughput is 1.6GB/s while bandwidth is 6.4GB/s	–	800 MH, 0.16 µm	–	–	-Up sampler delay is 7.5ns —Switch delay is 6ns —Down sampler delay is 11ns
18	Nostrum (Penolazzi and Jantsch 2006; Millberg et al. 2004)	2004	Packet size is 1 flit (128 bits)	Input and output buffer	–		–	–	–	Channel width is 128 bit	–
19	Nexus (Lines 2004)	2004	Burst word format is 36 bit data (plus 1 tail bit and 4 control bits)	–	–	–	-No physical architecture is provided for fault tolerance —Fault test pattern generated is for fault and delay testing	450MHz, 130nm	16	Channel width is 36 bits plus the tail bit and 4 bit To/ From channels	Latency is 2 to 3.1ns for various processes
20	Xpipes (Bertozzi and Benini 2004)	2004	Packet size is 64 byte (flit size is 32 bits)	Output queue (2 virtual channels)	OCP	2.2 GB/s	Distributed error detection bits	Multi-GHz, 0.1µm	4	–	Latency of a channel is more than one cycle

No.	Name	Year								
21	R²NoC (Samuelsson and Kumar 2004) (Ring Road NoC Architecture)	2004	–	–	–	–	–	3	–	–
22	An Architecture and Compiler for aSoC (Liang et al. 2004)	2004	Data stream	–	–	–	400 MHz, 0.18 µm	5	–	–
23	µSpider (Evain, Diguet, and Houzet 2004)	2004	Data width is 8 bit (configurable & generic number of virtual channels)	Input buffers (size is 3 words)	VCI	–	*No arbitration:* 96 MHz *Fixed channel priority:* 93 MHz *Round-robin priority:* 86 MHz	5 (configurable)	–	–
24	A 0.13µm NoC (Mondinelli, Borgatti, and KOVACS VAJNA 2004)	2004	–	Input & output FIFO's	–	1Gb/s/channel	200 MHz, 0.13µm	–	–	Average total latency in the worst case is 0.6µs
25	Spidergon (Coppola et al. 2004)	2004	–	–	–	–	–	–	–	–
26	RaSoC (Zeferino, Kreutz, and Susin 2004) (Router architecture for SoC)	2004	Flit size is n+2 (where n is 8, 16, 32 bits)	Input buffers (1 virtual channel)	–	–	55.8 MHz	5	–	Data bus width is 8, 16, 32 bits

(Continued)

SNo.	Architectures	Comparison Parameters									
		Year	Flit Size	Buffering	PE-Router Interface	Estimated Peak Performance	Fault Tolerance	NoC Frequency/ Technology	Switch/ Router Ports	Bus/ Channel width	Latency (per router/switch/ NoC)
27	QNoC (Bolotin et al. 2004) (Quality of Service NoC)	2004	16 bits	Input buffers (size of 2 flit)	–	*Signaling:* 320 Mbps *Real-Time:* 320 Mbps *RD/WR:* 2.56 Gbps *Block Transfer:* 2.56 Gbps	–	50 MHz	5	Bus width is 32 bits	–
28	On-Chip Network for Low Power Heterogeneous SoC Platform (Lee et al. 2004)	2004	Packet size is 80 bits	–	–	11.2 GB/s	–	1.6 GHz, 0.18μm	5 input ports and 3 output ports	–	Latency is 320ps per link
29	NoCGen (Chan and Parameswaran 2004)	2004	4, 8, 16 & 32 bits	Input & output buffer (buffer size is 8 to 32 flits)	–	-	–	–	–	Data bus width is 32 bits	–
30	Reconfigurable Network on Chip (Ching, Schaumont, and Verbauwhede 2004)	2004	Packet size is 32 bits	Internal buffering of 2 packets	–	1D network peak throughput is 1Gbis/sec	–	-1D network clock is 100 MHz —2D network clock is 85 MHz	3	Data bus width is 32 bits	Average round trip time from source to destination per hop is 17 cycles

31	DyAD (Hu and Marculescu 2004)	2004	Flits size is 32 bits	Input buffers (2 to 8 flits)	–	XY network saturates at 0.0167 packets/cycle while odd-even and DyAD saturates at 0.0256 and 0.027 packets/cycle	–	333 MHz, 0.16 μm	–	Inport link width is 32 bits	–
32	A Routing Switch for On-Chip Interconnection Networks (Chi and Chen 2004)	2004	Packet size is 256 bits	Input buffers (size is of 16 packets)	–	Saturation point of 0.5 (units not mentioned)	–	-100 MHz —150 MHz (hardware implementation, 0.25 μm)	5	Link width is 256 bits	–
33	Asynchronous On-Chip Network Router with Quality of Service (Feliciian and Furber 2004)	2004	-QoS2, 3 packet length is 5 flits —QoS1 packet is 134 to 1741 flits —BE packet length is 10 flits	-Input buffers (3 flits) —QoS1 requires output buffers 10 flits	–		–	Quasi-delay insensitive (QDI) technology	4	Data bus width is 8 bits	–
34	Black-Bus (Anjo et al. 2004)	2004	–	–	–	–	–	–	5	–	–
35	SNA (Lee, Lee, and Lee 2004) (SoC Network Architecture)	2004	–	–	AXI, AMBA AHB, VCI	–	–	–	–	Data bus width is 32 bits	Latency is 16 cycles & arbitration latency of AMBA AHB is 1 cycle.

(Continued)

SNo.	Architectures	Comparison Parameters									
		Year	Flit Size	Buffering	PE-Router Interface	Estimated Peak Performance	Fault Tolerance	NoC Frequency/ Technology	Switch/ Router Ports	Bus/ Channel width	Latency (per router/switch/ NoC)
36	Mango (Bjerregaard and Sparso 2004, 2005a, 2005b) (Message Passing Asynchronous Network on Chip providing Guaranteed services over OCP Interface)	2005	Packets are of variable length (flit size is 32 bit plus 5 bits for control)	Output buffers (one flit size)	OCP	702 MDI/s	–	515MHz, 0.12 µm	4	Bus width is 32bit	Flit time is 1.42ns (per switch)
37	Wolkotte et al. (Wolkotte et al. 2005)	2005	Packet size is 20 bits (16 data bits plus 4 header bits)	None	–	Bandwidth per link 17.2 Gbits/s	–	1075MHz, 0.13 µm	5	Bidirectional link is 16 bits wide	–
38	ASPIDA (Amde et al. 2005) network on chip	2005	*Master Packets:* Write: 10 bytes Read: 6 bytes *Slave Packets:* Write: 2 bytes Read: 6 bytes	–	–	–	–	0.18µm	–	–	–
39	Topology Adaptive NoC (Bartic et al. 2005)	2005	-Payload size is 544 bytes —Parameterize packet size	Output queue (one buffer of size 3 packets)	–	100 Mbyte/s per link	–	50 MHz	Parameterize	Channel width is 16 bits	–

		Year									
40	Spatial Division Multiplexing NoC (Leroy et al. 2005)	2005	–	–	–	640 Mbps per port	–	20 MHz, 130nm	5	-Link width is 8 bits —32-bit wide port	Critical path latency is 0.44ns
41	Arteris (Arteris 2005)	2005	32 bit	AHB, OCP, AXI	–	Greater than 100GB/s (for 36 initiators)	–	750 MHz, 90nm	5	–	-Gate delay is 60ps —Setup and hold time is 0.3ns —Propagation delay is 2ns —Wire transition takes 220ps —660ps to traverse a cluster —IP block latency is 6 clock cycles —For switch it takes 3 clock cycles —Average arbitration latency is 2 cycles
42	DyNoC (Bobda et al. 2005) (Dynamic NoC)	2005	–	–	–	–	–	75 MHz	–	Data bus width is 8 to 32 bits	Latency is 1 clock cycle
43	CDMA Router for On-Chip Switched Networks (Kim, Kim, and Sobelman 2005)	2005	–	–	Input & output buffers	-44.8Gbps (128 bit, 0.25μm) —83.7Gbps (128 bit, 0.18μm)	–	0.18 μm (94 MHz) & 0.25 μm (50 MHz)	7	–	Packet latency is 160ns
44	Asynchronous NoC Architecture (Beigné et al. 2005)	2005	Flit is 32 bit data plus 2 control bits	–	–	5 Gbytes/sec	–	250 MHz, 0.13 μm	5	Data bus width is 32 bits	–

(Continued)

SNo.	Architectures	Comparison Parameters										
		Year	Flit Size	Buffering	PE-Router Interface	Estimated Peak Performance	Fault Tolerance	NoC Frequency/ Technology	Switch/ Router Ports	Bus/ Channel width	Latency (per router/switch/ NoC)
45	Adaptive Network on Chip (Lee et al. 2005)	2005	–	–	–	1.2Gb/s	–	400 MHz, 0.18 μm	–	–	–
46	A Low Latency Router (Kim et al. 2005)	2005	Parametrizable flit size and number of flits per packet	-Virtual channels (Total numbers of VC=4 —4 flits per buffer per VC)	–	–	–	200 MHz, 70nm	5	–	–
47	Kavaldjiev et al. (Kavaldjiev et al. 2006)	2006	-A message is of size 256 Bytes —BE packet consists of 10 Byte payload	Input queue	–	*GT load*: —512 Mbits/s per node. —18.4Gbits/s for 36 nodes *BE load*: 640 Mbits/s	–	500MHz, 0.13 μm	–	Network channel is 16 bits	1.3 μsec is required for PE processing, receiving and transmitting respectively GT message latency is 1.242 μsec
48	DSPIN (Panades, Greiner, and Sheibanyrad 2006) (Distributed Scalable Predictable Interconnect Network)	2006	-Packet length is from 1 to 16 flits —32 bits for data and 2 bits for control	Input and output FIFO (BE: 8 flit GS: 4 flit)	-OCP —VCI	-The GS channel has a bandwidth of 8 Gbps —The BE bandwidth for 100 clusters is 400 Gbps	–	500MHz, 90nm	–	–	–

49	INoC (Neeb and Wehn 2008) (Irregular Network on Chip)	2006	Packet size is 4 and 32 filts	Input queue (size is 6 flits)	–	It achieves the saturation point of 0.40 to 0.60 for different type of traffic patterns and networks	–	–	–	–	–
50	XGFT (Kariniemi and Nurmi 2006) (Extended Generalized Fat Tree)	2006	Packet size is from eight to 64 flits (its width is 32 bits)	Input & output queue	–	–	Fault diagnosis and repair (FDAR) detects and repair static, dynamic and transient faults (bit errors and blocking of packets).	500 MHz, 130nm	–	Link width is 144 bits	–
51	High Throughput NoC Architecture (Bouhraoua and Elrabaa 2006)	2006	Packet size is 64 bytes & 128 bytes	-No buffers at router —Buffer at client(IP or core) interface (size is 64 bytes, 128 bytes, 256 bytes)	–	409.6Gbits/s	–	800 MHz, 0.13 microns	8	–	–
52	PNoC (Hilton and Nelson 2006) (Programmable NoC)	2006	–	Input & output buffers (parametrizable and optional)	–	–	–	124MHz	8	–	–

(Continued)

SNo.	Architectures	Comparison Parameters									
		Year	Flit Size	Buffering	PE-Router Interface	Estimated Peak Performance	Fault Tolerance	NoC Frequency/ Technology	Switch/ Router Ports	Bus/ Channel width	Latency (per router/switch/ NoC)
53	CTNOC (Sanusi and Wang 2006) (A Central Caching Network on Chip Communication Architecture Design)	2006	–	Input buffer	–	–	–	331.35 MHz	5 & control line	–	Latency is 3.018 ns
54	Low Latency on-Chip Network (Mullins, West, and Moore 2006)	2006	Packet length is 256 bits (4 flits)	Input buffers (4 VC's having size of 4 flits each)	–	16 Gbit/s	-Error detection code —Flits ordering checks	250 MHz to 300 MHz, 180nm	–	-Channel width is 80 bits —64 bit data & 16 bit control information	Latency is one cycle per hop
55	GEXPolygon & GEXSpidergon (Zid et al. 2006)	2006	Packet of size 3 flits (flit size is 32 bits)	–	–	–	-Cyclic Redundancy Code (CRC) for error detection —Faulty node/ interconnect detection but no detail provided	50 MHz	9, 6, 5 & 4	–	–

56	Low-Power Network on Chip (Lee, Lee, and Yoo 2006)	2006	-Packet size is 80 bits (16 bit header, 32 bit address, 32 bit data) —Burst size is 2, 4, 8 packets	Input buffer	–	11.2GB/s	–	1.6GHz, 0.18 μm	–	–	Latency is 0.9ns±40ps
57	CoNoChi (Pionteck, Koch, and Albrecht 2006)	2006	Payload size is 1024 bytes plus 12 bytes of header	–	–	–	–	73 MHz	–	Data bus width is 8, 16 & 32 bits	–
58	Cross Road Interconnection Architecture (Chang, Shen, and Chen 2006)	2006	–	–	OCP	–	–	0.18 μm	4	–	Latency per switch is 3.48 ns
59	HIBI (Salminen et al. 2006) (Heterogeneous IP Block Interconnection)	2006	–	Input & output FIFO's	OCP	–	–	-The frequency on ASIC is from 245 to 420 MHz —On Stratix FPGA frequency is from 81 MHz to 120 MHz. —Altera have frequency of 120 MHz -0.18 μm	5	32 bit wrapper	–
60	On-Chip Multimedia Applications (Lee et al. 2006)	2006	Flit size is 16 bits	Output buffers (size is 16 flits)	–	–	–	–	3 to 6	Network channel width is 16 bits	Latency is 4 clock cycle per router

(Continued)

SNo.	Architectures	Comparison Parameters									
		Year	Flit Size	Buffering	PE-Router Interface	Estimated Peak Performance	Fault Tolerance	NoC Frequency/ Technology	Switch/ Router Ports	Bus/ Channel width	Latency (per router/switch/ NoC)
61	Dynamic Reconfigurable NoC for Adaptive Reconfigurable MPSoC (Ahmad, Erdogan, and Khawam 2006)	2006	Packet size is 16 bits	–	–	–	Error detection and correction	–	5	Port link width of 32 bits	–
62	Communication Architecture Optimization (Ogras et al. 2006)	2006	Packet size is from 5 to 32 flits	–	–	0.40 packets/ cycle	–	100 MHz	5	–	4 clock cycles required to route the head flit
63	ProtoNoC (Castells-Rufas, Joven, and Carrabina 2006)	2006	Packet size is 32 bits	–	–	–	–	96 MHz	5	Channel width is 32 bits	–
64	TTNoC (Schoeberl 2007) (Time Triggered Network on Chip)	2007	–	NI contains 128X128 bit transmit and receive buffers	–	-Base bandwidth is 29 Gbits/s —230 Gbits/s for eight nodes.	–	-Network clock is 225MHz, —Individual nodes are clocked at 100MHz)	–	Bus width is 128bit	The minimum period when two nodes consumes the whole bandwidth is 4.5ns
65	NocMaker (Castells-Rufas et al. 2009)	2007	–	–	–	–	–	–	–	–	–

No.	Name	Year	Message/Packet	Buffer		Bandwidth		Frequency, Technology		Channel width	Latency
66	TILEPro64 (Bell et al. 2008)	2007	Message length is 1 to 128 flits (32 bits)	Input & output queue	–	240 GB/s bisection bandwidth	–	700–866 MHz, 90 nm	5	Data path is 32 bits	–
67	UT TRIPS (Gratz et al. 2007)	2007	Message length is 1 to 5 flits (138 bits)	Input & output queue (size is 2 flits)	–	-Peak injection bandwidth is 153 Gbytes/s —Bisection bandwidth is 44 Gbytes/s	–	366 MHz, 130 nm	6	Channel width is 138 bits	Router latency is 1 cycle
68	Intel TeraFLOPS (Vangal et al. 2008)	2007	-Packet size is 2 flits —Flit size is 38 bits (32 data bits plus 6 control bits)	Input buffers (size is 16 flits)	–	Peak bandwidth 2.56 terabits per second	–	4GHz, 65 nm	5	Channel width is 138bits	Per hop latency is 1nsec
69	Ambric (Butts 2007)	2007	Data bus is 32 bit	Input & output buffer	–	713Gbps	–	300 MHz, 130nm	–	32 bit bus	–
70	STNoC (Palermo et al. 2007)	2007	–	–	–	–	–	–	3	–	–
71	EIB on-Chip Network (Ainsworth and Pinkston 2007)	2007	Packet size is 128 bytes	–	–	In excess of 100 Gbytes/s (effective bandwidth)	–	–	–	–	–
72	SCC (Hoffman et al. 2007) (Scalable Communication Core)	2007	Flit size is 32 bits, Logical (data) packet size is 262144 bits	–	OCP	-Channel capacity is 8000 Mbps —Network capacity is 120Gbps	–	233 MHz, 90nm	5	–	–

(Continued)

SNo.	Architectures	Comparison Parameters									
		Year	Flit Size	Buffering	PE-Router Interface	Estimated Peak Performance	Fault Tolerance	NoC Frequency/ Technology	Switch/ Router Ports	Bus/ Channel width	Latency (per router/switch/ NoC)
73	Polaris (Soteriou et al. 2006)	2007	64 bit	Buffer size is 4, 16, 64 flits —virtual channels per links are 1,2, 4, 8	–	–	–	90nm (3GHz) & 50nm (10GHz)	5 & 9	–	Latency is 38.10 cycles to 151.12 cycles for different topologies
74	Generalized de Bruijn Graph NoC (Hosseinabady et al. 2007)	2007	–	Input and output buffers	–	640 Mbps	Reliable routing algorithm to detour the faulty interconnect	$0.18\mu m$, $0.13\mu m$, $0.09\mu m$	5	–	**-(ns, 0.18µm, VDD=1.8V)** *Normal switch*:6.79 *Reliable Switch*:6.79 **(ns, 0.13µm, VDD=1.2V)** *Normal switch*:3.27 *Reliable Switch*:3.37 **—(ns, 0.09µm, VDD=1V)** *Normal switch*:1.83 *Reliable Switch:* 1.96
75	MoCRes (Janarthanan, Swaminathan, and Tomko 2007)	2007	–	–	–	2.85 Gbit/s	–	Common clock: 286 MHz Multi-clock range from 357 MHz to 286 MHz	5	–	–

76	CSRA-NoC (Braun et al. 2007) (Circuit Switched Run Time Adaptive NoC)	2007	–	–	–	–	–	66 MHz	–	Data bus width is 8 bit	latency is 368 to 2448 clock cycles
77	artNoC (Schuck, Lamparth, and Becker 2007) (Adaptive Real-Time Network on Chip)	2007	Variable packet length	Virtual output queue (4 VC per port, 2 flit per VC)	–	40 Gbits/s	Fault tolerance using broadcasting of packets (concept not explained)	500 MHz, 0.18 μm	–	Data link width is 16 bits	Setup latency is 3 clock cycles
78	CDMA NoC (Wang, Ahonen, and Nurmi 2007)	2007	Packet cell is 32 bits	Buffers (32 bits)	VCI, OCP	–	–	0.18 μm	–	Data bus width is 32 bits	–
79	A Reconfigurable Baseband Platform Based on Asynchronous NoC (Lattard et al. 2008)	2008	Flit size is 34 bits (32 bit payload & 2 bit header)	–	–	5.120 Gbps per link	–	IP operating frequency is 162 MHz to 250 MHz 130 nm	5	–	Node latency is 6ns while network interface latency is 12ns
80	TTNoC (Paukovits and Kopetz 2008) (Time Triggered Network on Chip)	2008	Burst operation (contains one flit per cycle)	none	–	-Theoretical throughput achieved is 11.2Gbit/s —For 2 communication channel the throughput achieved is 22.4Gbits/s	Fault Tolerant (No detail provided)	Switch runs at 350MHz	4	Data bus width is 32bit	end-to-end latency is from nano seconds to seconds

(Continued)

SNo.	Architectures	Comparison Parameters									
		Year	Flit Size	Buffering	PE-Router Interface	Estimated Peak Performance	Fault Tolerance	NoC Frequency/ Technology	Switch/ Router Ports	Bus/ Channel width	Latency (per router/switch/ NoC)
81	EVC (Kumar et al. 2008) (Express Virtual Channels)	2008	–	Input buffers	–	–	–	–	5	–	–
82	ReNoC (Stensgaard and Sparsø 2008) Reconfigurable Network on Chip	2008	Packet size is 4 flits (96 bits) (Flit size is 34 bits having 32 data bits 2 control bits)	Input buffers (Having 2VC's, size of 4 flits)	–	2.4Gbits/s	–	100 MHz, 90nm	5	Link width is 37 bits	Latency of the link and topology switch is 120ps and 550ps
83	MoCSYS (Janarthanan and Tomko 2008) (Multi-Clock SYS)	2008	–	–	–	1912 MB/s	–	2.4 GHz	–	Data bus width is 32 bits	–
84	Aelite (Hansson, Subburaman, and Goossens 2009)	2009	3 words	Input queue (1 word size)	–	64Gbytes/s (an arity-6 aelite router for 64 bit data width)	–	800 MHz, 90 nm	Parametrizable ports —5 ports	32 bit data width	-Latency of connections is 35 to 500ns —Latency per hop is 3 cycles
85	HT-OCTAGON (El-Moursy, Korzec, and Ismail 2009) (High Throughput OCTAGON)	2009	8 wires for data and 4 for control between switches	Input and output buffers	–	Saturation point of 0.94 flit/ cycle/node for VC=8	–	500MHz	–	–	-Latency of 10 cycles per switch

86	XHiNoC (Samman, Hollstein, and Glesner 2009) (Extendable Hierarchical NoC)	2009	Packet size is 38 bits (32 data bits plus 6 control bits) —Packet contains 128 flits which include 1 header flit and 127 payload flits	Input FIFO (size is 8 registers)	–	37.76 Gbit/s	–	200 MHz, 130nm	5	–	–
87	ALPIN (Beigné et al. 2009) Asynchronous Low Power Innovative NoC	2009	32 bits	Input and output buffers	–	17Gbps	–	65 nm	5	–	–
88	Network on Chip in a Three Dimensional (Feero and Pande 2008)	2009	–	–	–	–	–	1.67 GHz, 90nm	–	Bus width is 128 bits	Wire delay of 600ps
89	BiNoC (Lan et al. 2011) (Bidirectional NoC)	2010	Packet size is 16 flits	Input buffers(2, 3, 4 VC's having 16 flits size)	–	Saturation point is from 0.3 flit/cycle/node to 0.7 flit/cycle/node	–	-In MHz 561, 627, 637, 666, 921, 90 nm	5	–	Latency is 4 clock cycles when a flit is moved between two routers (round trip)
90	RAMPSoC (Göhringer et al. 2010) (Run-time Adaptive Multiprocessor System on Chip)	2010	Control packet width is 15 bits	–	–	–	–	–	*Subswitch*: 4 *Superswitch*:9 *Root switch*:8	–	–

(*Continued*)

SNo.	Architectures	Comparison Parameters									
		Year	Flit Size	Buffering	PE-Router Interface	Estimated Peak Performance	Fault Tolerance	NoC Frequency/ Technology	Switch/ Router Ports	Bus/ Channel width	Latency (per router/switch/ NoC)
91	DRNoC (Krasteva, De la Torre, and Riesgo 2010) (Dynamic Reconfigurable NoC)	2010	–	Input buffers	AMBA APB	–	–	–	6	–	–
92	PMCNOC (Wang et al. 2010)	2010	–	Input buffer (No. of buffers=5, four VC's in each input port)	–	–	–	331.4 MHz to 335.91 MHz	5	–	Latency is from 2.977ns to 3.018 ns
93	Dynamic Reconfigurable Network on Chip (Wu, Tang, and Hsu 2011)	2011	Packet consists of 6 flits.	Input queue	–	-It achieves saturation point of 0.15 flit/ cycle/node for Transpose traffic. —It achieves saturation point of 0.17 flit/ cycle/node for Bit-Complement traffic.	–	–	–	–	–

94	Ramos et al. (Peña-Ramos and Parra-Michel 2011)	2011	Packet size is variable	—	—	*-Ad hoc:* 113MHz *—Internal Crossbar:* 113.08 MHz *—External Crossbar:* 113.359 MHz)	—	—	—
95	ORNoC (Le Beux et al. 2011) (Optical Ring Network on Chip)	2011	—	—	—	90nm	—	—	—
96	Kilo-NoC (Grot et al. 2011) (Thousand Connected NoC)	2011	Packet size is of four flits (Flit size is 16 byte)	Input buffer (Size of 70 flits)	—	32nm	—	Channel width is 128 bits	—
97	WiNoC (Ganguly et al. 2010)	2011	Packet length is 64 flits (flit size is 32 bits)	Input and output buffers (2 flit depth)	—	2.5 GHz, 65nm	—	Width of wired links are 32 bits	—
98	dAElite (Stefan, Molnos, and Goossens 2012) (distributed AElite)	2012	2 words & configuration word size of 7 bits	Input and output queue (16 word size) —32 slots	—	TSMC 65nm, 90nm, 130nm	3, 4, 5, 8 ports router	32 bit data width	-Latency per hop is 2 cycles —6 cycles for transmitting the slot table
99	BMNoC (Lee et al. 2012) (Busmesh Network on Chip)	2012	Packet consists of 32 flits. Each flit is 16 bits.	Input queue (32 flits size)	-200Mbits/s —It achieves saturation point of 0.35 flits/ cycle/node.	100MHz	—	•	—

(Continued)

SNo.	Architectures	Comparison Parameters										
		Year	Flit Size	Buffering	PE-Router Interface	Estimated Peak Performance	Fault Tolerance	NoC Frequency/ Technology	Switch/ Router Ports	Bus/ Channel width	Latency (per router/switch/ NoC)
100	Custom Network on Chip Architecture (Mishra, Nidhi, and Kishore 2012)	2012	Packet size is 6 bit (Four bit data plus 2 bit processors address)	No	–	–	–	–	•	–	–
101	DANoC (Shu et al. 2012) (A dynamic adaptive Network on Chip Architecture)	2012	–	Input buffer (Size of 4 flits)	–	-It achieves the saturation point of 0.17 to 0.32 for different injection rate	–	–	8	–	–
102	WaveSync (Yang et al. 2014)	2012	Packet length was varied from 2 to 5 flits	Input queue	–	-It achieved saturation point of 0.3 to 0.45 flit/node/ns for different traffic patterns	–	600MHz to 2000MHz, 45nm library	–	–	–

103	AdNoC (Al Faruque, Ebi, and Henkel 2010) (Adaptive Network on chip)	2012	8 bits	Input & output queue (size is 20 flits, 160 bits)	–	50Mbits/sec	-Detects faults during NoC adaptation at architectural level —No architectural detail provided. —No details about run time node and interconnect faults detection	1GHz	5	Link width is 8 bits	Critical path is 3.66ns
104	Aurora (Qouneh et al. 2012)	2012	Message size is 13312 bits	–	–	64 Gbps (Maximum transmission time is 208 cycles)	–	2GHz, 45 nm	–	–	–
105	Mesh-based NoC (Choudhary and Qureshi 2012)	2012	–	–	–	*Peak throughput*: 28Gbps (cross routing) 24 Gbps (fully adaptive)	–	–	5	–	*Latency clock cycles per flit*: 1.78 (cross routing) 2.13 (fully adaptive)

(*Continued*)

SNo.	Architectures	Comparison Parameters									
		Year	Flit Size	Buffering	PE-Router Interface	Estimated Peak Performance	Fault Tolerance	NoC Frequency/ Technology	Switch/ Router Ports	Bus/ Channel width	Latency (per router/switch/ NoC)
106	RecMIN (Logvinenko, Gremzow, and Tutsch 2013) (Reconfigurable Multi-Interconnection Network)	2013	Packet size is of 1 flit/phit	Input buffer (16 flit/phit)	–	0.63 flits per clock cycle	–	–	–	–	–
107	SWIFT (Postman et al. 2012) (Swing reduced Interconnect for a Token-based NoC)	2013	Message length is 5 flits (64 bits)	Input buffers (8 buffers per port)	–	5–9 Gbits/sec	–	400 MHz at low load, 225 MHz at high injection rate, 90 nm	–	–	–
108	SpiNNaker (Furber et al. 2012)	2013	40 bits or 72 bits	–	AMBA AHB	1Gbyte per second	–	100 MHz	6	–	0.1 µsec
109	H-NoC (Carrillo et al. 2012)	2013	48 bits or 56 bits	–	–	3.33G spikes per second	–	100 MHz, 65nm	4	–	-5.73 µsec —Delay between two cluster facilities is 660ns
110	BLOCON (Kao and Chao 2012) (Buffer less photonic CIOs Network)	2014	64 bits	Input queue at switch (Virtual output queue, size is 4 flits)	–	–	–	4 GHz, 32 nm	–	64 bit width for all links	Wire latency is 250ps

111	HELIX (Bahirat and Pasricha 2014) (Hydro nano photonic-electric network on chip)	2014	Flit width is 256 bits	Input and output queue (size of 4 flits)	–	40Gbps/link	–	3.8 GHz, 32nm	–	–	–
112	QORE (DiTomaso, Kodi, and Louri 2014)	2014	128 bits (packet size is four flits)	Input buffer (Quad function channel buffers, 6 buffers per link, 4 flit buffer)	–	-It achieved saturation point of 0.19 to 0.15 flit/node/cycle for 0% and 30% faults	-It uses buffers to overcome hard faults in the NoC.	2 GHz, 130 nm	20	–	The timing of our reversible channel buffer is 0.39 ns
113	DMesh (Wang and Bagherzadeh 2014)	2014	64 bits (packet length is 1, 8 and 16 flits)	Input buffer (buffer length is 2, 8 and 16 flits)	–	–	–	800 MHz, 65nm	11	64 bit	–
114	STNR (Poluri and Louri 2014)	2015	Packet length is 5 flits	Input buffer consists of four VCs (each VC can hold four flits)	–	–	-Detects and recovers from soft errors.	45 nm	5	–	The average latency has increased approximately by 0.5 percent
115	inoc (Heisswolf et al. 2015)	2015	32 bits	Input buffer	–	–	-Detects and recovers from transient, intermittent and permanent faults	45 nm	5	32 bit	–

(*Continued*)

SNo.	Architectures	Comparison Parameters									
		Year	Flit Size	Buffering	PE-Router Interface	Estimated Peak Performance	Fault Tolerance	NoC Frequency/ Technology	Switch/ Router Ports	Bus/ Channel width	Latency (per router/switch/ NoC)
116	TagNoC (Yaghini, Eghbal, and Bagherzadeh 2015)	2015	16 bits (packet size is 22 bytes)	Input buffer (128 bits)	–	-It achieved the saturation point of around 0.6 packets/node/ cycle for synthetic traffic —It achieved the saturation point of around 0.2 to 0.4 packets/node/ cycle under SPLASH-2 application traffic	–	40 nm	5	16 bit	–
117	Optical Multilevel Signaling (OMLS) NoC (Kao and Louri 2016)	2016	–	–	–	2.56 tera bits/sec	–	22 nm	5	–	–
118	Mesh Star Mesh (MSM) Architecture (Das and Ghosal 2017)	2017	32 bits	Input buffer (128 bits)	–	–	–	45 nm	–	–	–

119	Phosphorus Architecture (Mediouni and Hasnaoui 2017)	2017	32 bits	Input buffer (16 flits)	–	–	–	32 nm	3	–	–
120	DAPPER NoC Architecture (Raparti and Pasricha 2018)	2018	–	–	–	–	–	22 nm	5	128 bits	–
121	ShuttleNoC Architecture (Lu et al. 2018)	2019	–	Input buffer (5 flits depth)	–	–	–	45 nm	–	–	–
122	TAONoC Architecture (Yang et al. 2018)	2019	4096 bits (packet length)	–	–	3200 Gbps (ideal bandwidth)	–	–	–	–	–

REFERENCES

Ahmad, Balal, Ahmet T Erdogan, and Sami Khawam. 2006. "Architecture of a dynamically reconfigurable NoC for adaptive reconfigurable MPSoC." *First NASA/ESA Conference on Adaptive Hardware and Systems (AHS'06)*.

Ainsworth, Thomas William, and Timothy Mark Pinkston. 2007. "Characterizing the Cell EIB on-chip network." *IEEE Micro* 27 (5):6–14.

Al Faruque, Mohammad Abdullah, Thomas Ebi, and Jörg Henkel. 2010. "AdNoC: Runtime adaptive network-on-chip architecture." *IEEE Transactions on Very Large Scale Integration (VLSI) Systems* 20 (2):257–269.

Amde, Manish, Tomaz Felicijan, Aristeidis Efthymiou, Douglas Edwards, and Luciano Lavagno. 2005. "Asynchronous on-chip networks." *IEE Proceedings-Computers and Digital Techniques* 152 (2):273–283.

Anjo, Kenichiro, Yutaka Yamada, Michihiro Koibuchi, Akiya Jouraku, and Hideharu Amano. 2004. "BLACK-BUS: A new data-transfer technique using local address on networks-on-chips." *2004 18th International Parallel and Distributed Processing Symposium. Proceedings*.

Arteris, SA. 2005. "A comparison of network-on-chip and busses." *White Paper*.

Bahirat, Shirish, and Sudeep Pasricha. 2014. "HELIX: Design and synthesis of hybrid nano-photonic application-specific network-on-chip architectures." *Fifteenth International Symposium on Quality Electronic Design*.

Bainbridge, John, and Steve Furber. 2002. "Chain: A delay-insensitive chip area interconnect." *IEEE Micro* (5):16–23.

Bartic, TA, J-Y Mignolet, Vincent Nollet, Theodore Marescaux, Diederik Verkest, Serge Vernalde, and Rudy Lauwereins. 2005. "Topology adaptive network-on-chip design and implementation." *IEE Proceedings-Computers and Digital Techniques* 152 (4):467–472.

Beigné, Edith, Fabien Clermidy, Hélène Lhermet, Sylvain Miermont, Yvain Thonnart, Xuan-Tu Tran, Alexandre Valentian, Didier Varreau, Pascal Vivet, and Xavier Popon. 2009. "An asynchronous power aware and adaptive NoC based circuit." *IEEE Journal of Solid-State Circuits* 44 (4):1167–1177.

Beigné, Edith, Fabien Clermidy, Pascal Vivet, Alain Clouard, and Marc Renaudin. 2005. "An asynchronous NoC architecture providing low latency service and its multi-level design framework." *11th IEEE International Symposium on Asynchronous Circuits and Systems*.

Bell, Shane, Bruce Edwards, John Amann, Rich Conlin, Kevin Joyce, Vince Leung, John MacKay, Mike Reif, Liewei Bao, and John Brown. 2008. "Tile64-processor: A 64-core soc with mesh interconnect." *2008 IEEE International Solid-State Circuits Conference-Digest of Technical Papers*.

Bertozzi, Davide, and Luca Benini. 2004. "Xpipes: A network-on-chip architecture for gigas-cale systems-on-chip." *IEEE Circuits and Systems Magazine* 4 (2):18–31.

Bjerregaard, Tobias, and Jens Sparsø. 2004. "Virtual channel designs for guaranteeing band-width in asynchronous network-on-chip." *2004 Proceedings Norchip Conference*.

Bjerregaard, Tobias, and Jens Sparsø. 2005a. "A router architecture for connection-oriented service guarantees in the MANGO clockless network-on-chip." *Design, Automation and Test in Europe*.

Bjerregaard, Tobias, and Jens Sparso. 2005b. "Scheduling discipline for latency and bandwidth guarantees in asynchronous network-on-chip." *11th IEEE International Symposium on Asynchronous Circuits and Systems*.

Bobda, Christophe, Ali Ahmadinia, Mateusz Majer, Jürgen Teich, Sándor Fekete, and Jan van der Veen. 2005. "Dynoc: A dynamic infrastructure for communication in dynamically reconfugurable devices." *2005 International Conference on Field Programmable Logic and Applications*.

Appendix A

Bolotin, Evgeny, Israel Cidon, Ran Ginosar, and Avinoam Kolodny. 2004. "QNoC: QoS architecture and design process for network on chip." *Journal of Systems Architecture* 50 (2–3):105–128.

Bouhraoua, A, and ME Elrabaa. 2006. "A high-throughput network-on-chip architecture for systems-on-chip interconnect." *2006 International Symposium on System-on-Chip*.

Braun, Lars, Michael Hubner, Jurgen Becker, Thomas Perschke, Volker Schatz, and Stefan Bach. 2007. "Circuit switched run-time adaptive network-on-chip for image processing applications." *2007 International Conference on Field Programmable Logic and Applications*.

Butts, Mike. 2007. "Synchronization through communication in a massively parallel processor array." *IEEE Micro* 27 (5):32–40.

Carrillo, Snaider, Jim Harkin, Liam J McDaid, Fearghal Morgan, Sandeep Pande, Seamus Cawley, and Brian McGinley. 2012. "Scalable hierarchical network-on-chip architecture for spiking neural network hardware implementations." *IEEE Transactions on Parallel and Distributed Systems* 24 (12):2451–2461.

Castells-Rufas, David, Jaume Joven, and Jordi Carrabina. 2006. "A validation and performance evaluation tool for ProtoNoC." *2006 International Symposium on System-on-Chip*.

Castells-Rufas, David, Jaume Joven, Sergi Risueño, Eduard Fernandez, and Jordi Carrabina. 2009. "NocMaker: A cross-platform open-source design space exploration tool for networks on chip." *INA-OCMC Workshop*, Paphos, Cyprus.

Chan, Jeremy, and Sri Parameswaran. 2004. "NoCGEN: A template based reuse methodology for networks on chip architecture." *17th International Conference on VLSI Design. Proceedings*.

Chang, Kuei-Chung, Jih-Sheng Shen, and Tien-Fu Chen. 2006. "Evaluation and design trade-offs between circuit-switched and packet-switched NOCs for application-specific SOCs." *Proceedings of the 43rd Annual Design Automation Conference*.

Chi, Hsin-Chou, and Jia-Hung Chen. 2004. "Design and implementation of a routing switch for on-chip interconnection networks." *Proceedings of 2004 IEEE Asia-Pacific Conference on Advanced System Integrated Circuits*.

Ching, Doris, Patrick Schaumont, and Ingrid Verbauwhede. 2004. "Integrated modeling and generation of a reconfigurable network-on-chip." *2004 18th International Parallel and Distributed Processing Symposium. Proceedings*.

Choudhary, Sudhanshu, and Shafi Qureshi. 2012. "Performance evaluation of mesh-based NoCs: Implementation of a new architecture and routing algorithm." *International Journal of Automation and Computing* 9 (4):403–413.

Coppola, Marcello, Riccardo Locatelli, Giuseppe Maruccia, Lorenzo Pieralisi, and Alberto Scandurra. 2004. "Spidergon: A novel on-chip communication network." *2004 International Symposium on System-on-Chip. Proceedings*.

Dally, William J, and Brian Towles. 2001. "Route packets, not wires: On-chip interconnection networks." *Proceedings of the 38th Annual Design Automation Conference*.

Das, Tuhin Subhra, and Prasun Ghosal. 2017. "MSM: Performance Enhancing Area and Congestion Aware Network-on-Chip Architecture." *2017 IEEE International Symposium on Nanoelectronic and Information Systems (iNIS)*.

DiTomaso, Dominic, Avinash Kodi, and Ahmed Louri. 2014. "QORE: A fault tolerant network-on-chip architecture with power-efficient quad-function channel (QFC) buffers." *2014 IEEE 20th International Symposium on High Performance Computer Architecture (HPCA)*.

El-Moursy, Magdy A, Darek Korzec, and Mohammed Ismail. 2009. "High throughput architecture for OCTAGON network on chip." *2009 16th IEEE International Conference on Electronics, Circuits and Systems-(ICECS 2009)*.

Evain, Samuel, J-P Diguet, and Dominique Houzet. 2004. "μ spider: A CAD tool for efficient NoC design." *2004 Proceedings Norchip Conference*.

Feero, Brett Stanley, and Partha Pratim Pande. 2008. "Networks-on-chip in a three-dimensional environment: A performance evaluation." *IEEE Transactions on Computers* 58 (1):32–45.

Feliciian, F, and Stephen B Furber. 2004. "An asynchronous on-chip network router with quality-of-service (QoS) support." *2004 IEEE International SOC Conference. Proceedings.*

Forsell, Martti. 2002. "A scalable high-performance computing solution for networks on chips." *IEEE Micro* 22 (5):46–55.

Furber, Steve B, David R Lester, Luis A Plana, Jim D Garside, Eustace Painkras, Steve Temple, and Andrew D Brown. 2012. "Overview of the SpiNNaker system architecture." *IEEE Transactions on Computers* 62 (12):2454–2467.

Ganguly, Amlan, Kevin Chang, Sujay Deb, Partha Pratim Pande, Benjamin Belzer, and Christof Teuscher. 2010. "Scalable hybrid wireless network-on-chip architectures for multicore systems." *IEEE Transactions on Computers* 60 (10):1485–1502.

Göhringer, Diana, Michael Hübner, Laure Hugot-Derville, and Jürgen Becker. 2010. "Message passing interface support for the runtime adaptive multi-processor system-on-chip RAMPSoC." *2010 International Conference on Embedded Computer Systems: Architectures, Modeling and Simulation.*

Gratz, Paul, Changkyu Kim, Karthikeyan Sankaralingam, Heather Hanson, Premkishore Shivakumar, Stephen W Keckler, and Doug Burger. 2007. "On-chip interconnection networks of the TRIPS chip." *IEEE Micro* 27 (5):41–50.

Grot, Boris, Joel Hestness, Stephen W Keckler, and Onur Mutlu. 2011. "Kilo-NoC: A heterogeneous network-on-chip architecture for scalability and service guarantees." *ACM SIGARCH Computer Architecture News* 39 (3):401–412.

Guerrier, Pierre, and Alain Greiner. 2008. "A generic architecture for on-chip packet-switched interconnections." *Design, Automation, and Test in Europe.*

Hansson, Andreas, Mahesh Subburaman, and Kees Goossens. 2009. "Aelite: A flit-synchronous network on chip with composable and predictable services." *Proceedings of the Conference on Design, Automation and Test in Europe.*

Heisswolf, Jan, Andreas Weichslgartner, Aurang Zaib, Stephanie Friederich, Leonard Masing, Carsten Stein, Marco Duden, Roman Klöpfer, Jürgen Teich, and Thomas Wild. 2015. "Fault-tolerant communication in invasive networks on chip." *2015 NASA/ESA Conference on Adaptive Hardware and Systems (AHS).*

Henriksson, Tomas, Daniel Wiklund, and Dake Liu. 2003. "VLSI implementation of a switch for on-chip networks." *Proceedings of the International Workshop on Design and Diagnostics of Electronic Circuits and Systems,* Poznan, Poland.

Hilton, Clint, and Brent Nelson. 2006. "PNoC: A flexible circuit-switched NoC for FPGA-based systems." *IEE Proceedings-Computers and Digital Techniques* 153 (3):181–188.

Hoffman, Jeff, David Arditti Ilitzky, Anthony Chun, and Aliaksei Chapyzhenka. 2007. "Architecture of the scalable communications core." *First International Symposium on Networks-on-Chip (NOCS'07).*

Hosseinabady, Mohammad, Mohammad Reza Kakoee, Jimson Mathew, and Dhiraj K Pradhan. 2007. "Reliable network-on-chip based on generalized de Bruijn graph." *2007 IEEE International High Level Design Validation and Test Workshop.*

Hu, Jingcao, and Radu Marculescu. 2004. "DyAD: Smart routing for networks-on-chip." *Proceedings of the 41st Annual Design Automation Conference.*

Janarthanan, Arun, Vijay Swaminathan, and Karen A Tomko. 2007. "MoCReS: An area-efficient multi-clock on-chip network for reconfigurable systems." *IEEE Computer Society Annual Symposium on VLSI (ISVLSI'07).*

Janarthanan, Arun, and Karen A Tomko. 2008. "MoCSYS: A multi-clock hybrid two-layer router architecture and integrated topology synthesis framework for system-level design of FPGA based on-chip networks." *21st International Conference on VLSI Design (VLSID 2008).*

Appendix A 179

Kao, Tzyy-Juin, and Ahmed Louri. 2016. "Design of high bandwidth photonic NoC architectures using optical multilevel signaling." *2016 Tenth IEEE/ACM International Symposium on Networks-on-Chip (NOCS)*.

Kao, Yu-Hsiang, and H Jonathan Chao. 2012. "Design of a bufferless photonic clos network-on-chip architecture." *IEEE Transactions on Computers* 63 (3):764–776.

Karim, Faraydon, Anh Nguyen, and Sujit Dey. 2002. "An interconnect architecture for networking systems on chips." *IEEE Micro* 22 (5):36–45.

Kariniemi, Heikki, and Jari Nurmi. 2006. "On-line reconfigurable XGFT network-on-chip designed for improving the fault-tolerance and manufacturability of the MPSoC chips." *2006 International Conference on Field Programmable Logic and Applications*.

Kavaldjiev, Nikolay, Gerard JM Smit, Pierre G Jansen, and Pascal T Wolkotte. 2006. "A virtual channel network-on-chip for GT and BE traffic." *IEEE Computer Society Annual Symposium on Emerging VLSI Technologies and Architectures (ISVLSI'06)*.

Kim, Daewook, Manho Kim, and Gerald E Sobelman. 2005. "Design of a high-performance scalable CDMA router for on-chip switched networks." 대한전자공학회 *ISOCC*:32–35.

Kim, Jongman, Dongkook Park, Theo Theocharides, Narayanan Vijaykrishnan, and Chita R Das. 2005. "A low latency router supporting adaptivity for on-chip interconnects." *2005 42nd Design Automation Conference. Proceedings*.

Krasteva, Yana E, Eduardo De la Torre, and Teresa Riesgo. 2010. "Reconfigurable networks on chip: DRNoC architecture." *Journal of Systems Architecture* 56 (7):293–302.

Kumar, Amit, Li-Shiuan Peh, Partha Kundu, and Niraj K Jha. 2008. "Toward ideal on-chip communication using express virtual channels." *IEEE Micro* 28 (1):80–90.

Kumar, Shashi, Axel Jantsch, J-P Soininen, Martti Forsell, Mikael Millberg, Johny Oberg, Kari Tiensyrja, and Ahmed Hemani. 2002. "A network on chip architecture and design methodology." *Proceedings IEEE Computer Society Annual Symposium on VLSI. New Paradigms for VLSI Systems Design (ISVLSI 2002)*.

Lan, Ying-Cherng, Hsiao-An Lin, Shih-Hsin Lo, Yu Hen Hu, and Sao-Jie Chen. 2011. "A bidirectional NoC (BiNoC) architecture with dynamic self-reconfigurable channel." *IEEE Transactions on Computer-Aided Design of Integrated Circuits and Systems* 30 (3):427–440.

Lattard, Didier, Edith Beigne, Fabien Clermidy, Yves Durand, Romain Lemaire, Pascal Vivet, and Friedbert Berens. 2008. "A reconfigurable baseband platform based on an asynchronous network-on-chip." *IEEE Journal of Solid-State Circuits* 43 (1):223–235.

Le Beux, Sébastien, Jelena Trajkovic, Ian O'Connor, Gabriela Nicolescu, Guy Bois, and Pierre Paulin. 2011. "Optical ring network-on-chip (ORNoC): Architecture and design methodology." *2011 Design, Automation & Test in Europe*.

Lee, Hyung Gyu, Umit Y Ogras, Radu Marculescu, and Naehyuck Chang. 2006. "Design space exploration and prototyping for on-chip multimedia applications." *Proceedings of the 43rd Annual Design Automation Conference*.

Lee, Kangmin, Se-Joong Lee, Sung-Eun Kim, Hye-Mi Choi, Donghyun Kim, Sunyoung Kim, Min-Wuk Lee, and Hoi-Jun Yoo. 2004. "A 51mW 1.6 GHz on-chip network for low-power heterogeneous SoC platform." *2004 IEEE International Solid-State Circuits Conference (IEEE Cat. No. 04CH37519)*.

Lee, Kangmin, Se-Joong Lee, and Hoi-Jun Yoo. 2006. "Low-power network-on-chip for high-performance SoC design." *IEEE Transactions on Very Large Scale Integration (VLSI) Systems* 14 (2):148–160.

Lee, Sanghun, Chanho Lee, and Hyuk-Jae Lee. 2004. "A new multi-channel on-chip-bus architecture for system-on-chips." *2004 IEEE International SOC Conference. Proceedings*.

Lee, Se-Joong, Kwanho Kim, Hyejung Kim, Namjun Cho, and Hoi-Jun Yoo. 2005. "Adaptive network-on-chip with wave-front train serialization scheme." *Digest of Technical Papers. 2005 Symposium on VLSI Circuits.*

Lee, Se-Joong, Seong-Jun Song, Kangmin Lee, Jeong-Ho Woo, Sung-Eun Kim, Byeong-Gyu Nam, and Hoi-Jun Yoo. 2003. "An 800MHz star-connected on-chip network for application to systems on a chip." *2003 IEEE International Solid-State Circuits Conference. Digest of Technical Papers. ISSCC.*

Lee, Seungju, Nozomu Togawa, Yusuke Sekihara, Takashi Aoki, and Akira Onozawa. 2012. "A hybrid NoC architecture utilizing packet transmission priority control method." *2012 IEEE Asia Pacific Conference on Circuits and Systems.*

Leroy, Anthony, Paul Marchal, Adelina Shickova, Francky Catthoor, Frédéric Robert, and Diederik Verkest. 2005. "Spatial division multiplexing: A novel approach for guaranteed throughput on NoCs." *Proceedings of the 3rd IEEE/ACM/IFIP International Conference on Hardware/software Codesign and System Synthesis.*

Liang, Jian, Andrew Laffely, Sriram Srinivasan, and Russell Tessier. 2004. "An architecture and compiler for scalable on-chip communication." *IEEE Transactions on Very Large Scale Integration (VLSI) Systems* 12 (7):711–726.

Liang, Jian, Sriram Swaminathan, and Russell Tessier. 2000. "ASOC: A scalable, single-chip communications architecture." *Proceedings 2000 International Conference on Parallel Architectures and Compilation Techniques (Cat. No. PR00622).*

Lines, Andrew. 2004. "Asynchronous interconnect for synchronous SoC design." *IEEE Micro* 24 (1):32–41.

Logvinenko, Alexander, Carsten Gremzow, and Dietmar Tutsch. 2013. "RecMIN: A reconfiguration architecture for network on chip." *2013 8th International Workshop on Reconfigurable and Communication-Centric Systems-on-Chip (ReCoSoC).*

Lu, Hang, Yisong Chang, Guihai Yan, Ning Lin, Xin Wei, and Xiaowei Li. 2018. "ShuttleNoC: Power-adaptable communication infrastructure for many-core processors." *IEEE Transactions on Computer-Aided Design of Integrated Circuits and Systems.*

Mediouni, Nejib, and Salem Hasnaoui. 2017. "Phosphorus: An ultra low footprint and energy consumption 3D NoC architecture." *2017 International Conference on Internet of Things, Embedded Systems and Communications (IINTEC).*

Millberg, Mikael, Erland Nilsson, Rikard Thid, and Axel Jantsch. 2004. "Guaranteed bandwidth using looped containers in temporally disjoint networks within the Nostrum network on chip." *Proceedings Design, Automation and Test in Europe Conference and Exhibition.*

Mishra, Prabhakar, A Nidhi, and JK Kishore. 2012. "Custom Network on Chip architecture for map generation in autonomous navigating robots." *2012 Annual IEEE India Conference (INDICON).*

Mondinelli, Filippo, Michele Borgatti, and Zsolt M Kovacs Vajna. 2004. "A 0.13 um 1Gb/s/ channel store-and-forward network on-chip." *IEEE Systems-on-Chip Conference.*

Moraes, Fernando, Ney Calazans, Aline Mello, Leandro Möller, and Luciano Ost. 2004. "HERMES: An infrastructure for low area overhead packet-switching networks on chip." *INTEGRATION, the VLSI Journal* 38 (1):69–93.

Mullins, Robert, Andrew West, and Simon Moore. 2006. "The design and implementation of a low-latency on-chip network." *Proceedings of the 2006 Asia and South Pacific Design Automation Conference.*

Neeb, Christian, and Norbert Wehn. 2008. "Designing efficient irregular networks for heterogeneous systems-on-chip." *Journal of Systems Architecture* 54 (3–4):384–396.

Ogras, Umit Y, Radu Marculescu, Hyung Gyu Lee, and Naehyuck Chang. 2006. "Communication architecture optimization: Making the shortest path shorter in regular networks-on-chip." *Proceedings of the Conference on Design, Automation and Test in Europe. Proceedings.*

Appendix A 181

Palermo, Gianluca, Cristina Silvano, Giovanni Mariani, Riccardo Locatelli, and Marcello Coppola. 2007. "Application-specific topology design customization for STNoC." *10th EUROMICRO Conference on Digital System Design Architectures, Methods and Tools (DSD 2007)*.

Panades, Ivan Miro, Alain Greiner, and Abbas Sheibanyrad. 2006. "A low cost network-on-chip with guaranteed service well suited to the GALS approach." *2006 1st International Conference on Nano-Networks and Workshops*.

Pande, Partha Pratim, Cristian Grecu, André Ivanov, and Res Saleh. 2003. "High-throughput switch-based interconnect for future SoCs." *2003 3rd IEEE International Workshop on System-on-Chip for Real-Time Applications. Proceedings*.

Paukovits, Christian, and Hermann Kopetz. 2008. "Concepts of switching in the time-triggered network-on-chip." *2008 14th IEEE International Conference on Embedded and Real-Time Computing Systems and Applications*.

Peña-Ramos, JC, and Ramon Parra-Michel. 2011. "Network on chip architectures for high performance digital signal processing using a configurable core." *2011 International Conference on Reconfigurable Computing and FPGAs*.

Penolazzi, Sandro, and Axel Jantsch. 2006. "A high level power model for the Nostrum NoC." *9th EUROMICRO Conference on Digital System Design (DSD'06)*.

Pionteck, Thilo, Roman Koch, and Carsten Albrecht. 2006. "Applying partial reconfiguration to networks-on-chips." *2006 International Conference on Field Programmable Logic and Applications*.

Poluri, Pavan, and Ahmed Louri. 2014. "A soft error tolerant network-on-chip router pipeline for multi-core systems." *IEEE Computer Architecture Letters* 14 (2):107–110.

Postman, Jacob, Tushar Krishna, Christopher Edmonds, Li-Shiuan Peh, and Patrick Chiang. 2012. "Swift: A low-power network-on-chip implementing the token flow control router architecture with swing-reduced interconnects." *IEEE Transactions on Very Large Scale Integration (VLSI) Systems* 21 (8):1432–1446.

Qouneh, Amer, Zhongqi Li, Madhura Joshi, Wangyuan Zhang, Xin Fu, and Tao Li. 2012. "Aurora: A thermally resilient photonic network-on-chip architecture." *2012 IEEE 30th International Conference on Computer Design (ICCD)*.

Raparti, Venkata Yaswanth, and Sudeep Pasricha. 2018. "DAPPER: Data aware approximate NoC for GPGPU architectures." *Proceedings of the Twelfth IEEE/ACM International Symposium on Networks-on-Chip*.

Rijpkema, Edwin, Kees Goossens, Andrei Rădulescu, John Dielissen, Jef van Meerbergen, Paul Wielage, and Erwin Waterlander. 2003. "Trade-offs in the design of a router with both guaranteed and best-effort services for networks on chip." *IEE Proceedings-Computers and Digital Techniques* 150 (5):294–302.

Salminen, Erno, Tero Kangas, Timo D Hämäläinen, Jouni Riihimäki, Vesa Lahtinen, and Kimmo Kuusilinna. 2006. "HIBI communication network for system-on-chip." *Journal of VLSI Signal Processing Systems for Signal, Image and Video Technology* 43 (2–3):185–205.

Samman, Faizal A, Thomas Hollstein, and Manfred Glesner. 2009. "Networks-on-chip based on dynamic wormhole packet identity mapping management." *VLSI Design*:2.

Samuelsson, Henrik, and Shashi Kumar. 2004. "Ring road NoC architecture." *2004 Proceedings Norchip Conference*.

Sanusi, Azeez, and Magdy A Bayoumi Nan Wang. 2006. "A central caching network-on-chip communication architecture design." https://www.design-reuse.com/articles/15634/a-central-caching-network-on-chip-communication-architecture-design.html

Schoeberl, Martin. 2007. "A time-triggered network-on-chip." *2007 International Conference on Field Programmable Logic and Applications*.

Schuck, Christian, Stefan Lamparth, and Jurgen Becker. 2007. "artNoC-A novel multi-functional router architecture for Organic Computing." *2007 International Conference on Field Programmable Logic and Applications*.

Shu, Hao, Jiang-Yi Shi, Yue Hao, Pei-Jun Ma, and Zhao Xu. 2012. "DANoC: A dynamic adaptive network on chip architecture." *2012 IEEE 11th International Conference on Solid-State and Integrated Circuit Technology.*

Siguenza-Tortosa, D, and Jari Nurmi. 2002. "Proteo: A new approach to network-on-chip." *Proceedings, IASTED-Communication Systems and Networks (CSN 2002).*

Soteriou, Vassos, Noel Eisley, Hangsheng Wang, Bin Li, and Li-Shiuan Peh. 2006. "Polaris: A system-level roadmap for on-chip interconnection networks." *2006 International Conference on Computer Design.*

Stefan, Radu Andrei, Anca Molnos, and Kees Goossens. 2012. "dAElite: A TDM NoC supporting QoS, multicast, and fast connection set-up." *IEEE Transactions on Computers* 63 (3):583–594.

Stensgaard, Mikkel Bystrup, and Jens Sparsø. 2008. "ReNoC: A network-on-chip architecture with reconfigurable topology." *Proceedings of the Second ACM/IEEE International Symposium on Networks-on-Chip.*

Taylor, Michael Bedford, Jason Kim, Jason Miller, David Wentzlaff, Fae Ghodrat, Ben Greenwald, Henry Hoffman, Paul Johnson, Jae-Wook Lee, and Walter Lee. 2002. "The raw microprocessor: A computational fabric for software circuits and general-purpose programs." *IEEE Micro* 22 (2):25–35.

Vangal, Sriram R, Jason Howard, Gregory Ruhl, Saurabh Dighe, Howard Wilson, James Tschanz, David Finan, Arvind Singh, Tiju Jacob, and Shailendra Jain. 2008. "An 80-tile sub-100-w teraflops processor in 65-nm CMOS." *IEEE Journal of Solid-State Circuits* 43 (1):29–41.

Wang, Chifeng, and Nader Bagherzadeh. 2014. "Design and evaluation of a high throughput qos-aware and congestion-aware router architecture for network-on-chip." *Microprocessors and Microsystems* 38 (4):304–315.

Wang, Nan, Azeez Sanusi, PY Zhao, M Elgamel, and Magdy A Bayoumi. 2010. "PMCNOC: A pipelining multi-channel central caching network-on-chip communication architecture design." *Journal of Signal Processing Systems* 60 (3):315–331.

Wang, Xin, Tapani Ahonen, and Jari Nurmi. 2007. "Applying CDMA technique to network-on-chip." *IEEE Transactions on Very Large Scale Integration (VLSI) Systems* 15 (10):1091–1100.

Wiklund, Daniel, and Dake Liu. 2003. "SoCBUS: Switched network on chip for hard real time embedded systems." *Proceedings International Parallel and Distributed Processing Symposium.*

Wingard, Drew. 2001. "MicroNetwork-based integration for SOCs." *Proceedings of the 38th Design Automation Conference (IEEE Cat. No. 01CH37232).*

Wolkotte, Pascal T, Gerard JM Smit, Gerard K Rauwerda, and Lodewijk T Smit. 2005. "An energy-efficient reconfigurable circuit-switched network-on-chip." *19th IEEE International Parallel and Distributed Processing Symposium.*

Wu, Li-Wei, Wei-Xiang Tang, and Yarsun Hsu. 2011. "A novel architecture and routing algorithm for dynamic reconfigurable network-on-chip." *2011 IEEE Ninth International Symposium on Parallel and Distributed Processing with Applications.*

Yaghini, Pooria M, Ashkan Eghbal, and Nader Bagherzadeh. 2015. "On the design of hybrid routing mechanism for mesh-based network-on-chip." *Integration* 50:183–192.

Yang, Yintang, Ke Chen, Huaxi Gu, Bowen Zhang, and Lijing Zhu. 2018. "TAONoC: A regular passive optical network-on-chip architecture based on comb switches." *IEEE Transactions on Very Large Scale Integration (VLSI) Systems* 27 (4):954–963.

Yang, Yoon Seok, Reeshav Kumar, Gwan Choi, and Paul V Gratz. 2014. "WaveSync: Low-latency source-synchronous bypass network-on-chip architecture." *ACM Transactions on Design Automation of Electronic Systems (TODAES)* 19 (4):34.

Zeferino, Cesar Albenes, Márcio Eduardo Kreutz, and Altamiro Amadeu Susin. 2004. "RASoC: A router soft-core for networks-on-chip." *Proceedings of the Conference on Design, Automation and Test in Europe-Volume 3.*

Appendix A

Zeferino, Cesar Albenes, and Altamiro Amadeu Susin. 2003. "SoCIN: A parametric and scalable network-on-chip." *2003 16th Symposium on Integrated Circuits and Systems Design. SBCCI 2003. Proceedings.*

Zid, Mounir, Abdelkrim Zitouni, Adel Baganne, and Rached Tourki. 2006. "New generic GALS NoC architectures with multiple QoS." *2006 International Conference on Design and Test of Integrated Systems in Nanoscale Technology (DTIS 2006).*

Appendix B

Bio-Inspired NoC Fault-Tolerant Codes

Synaptogenesis Algorithm

```
if (portNumber==0)
{
        int northPort=curRouter->par("northPort_W");
        if (northPort==0)
          {
            faultySynpseID=synpseID;
            msg->setLostFlit(true);
            send(msg, "sched_link$o");
            creditPop=new NoCPopCreditMsg();
            creditPop->setKind(NOC_POPCredit_Msg);
            creditPop->setVc(msg->getVC());
            scheduleAt(simTime()+0.000000002,creditPop);
            return;
          }
        else
          {

//EV<<"Checking again if port is working, whether the flit
belong to faultySynapseID, if yes, send it to inPort to send
it on new workaround synapse"<<endl;

        if (faultySynpseID==synpseID)
          {
            msg->setLostFlit(true);
            send(msg, "sched_link$o");
            creditPop=new NoCPopCreditMsg();
            creditPop->setKind(NOC_POPCredit_Msg);
            creditPop->setVc(msg->getVC());
            scheduleAt(simTime()+0.000000002,creditPop);
            return;
          }
        }
      }

//EV<<"Making a copy of the sent flit, may be port become
faulty and credit message is not received from neighbor
router"<<endl;

                copyFlit=msg->dup();
                qSched.insert(copyFlit);
```

185

186 Appendix B

```
//EV<<"Retrieving data from flit, router/interconnect might be-
come faulty"<<endl;

        int type=msg->getType();
        int vc=msg->getVC();
        int srcId=msg->getSrcId();
        int dstId=msg->getDstId();
        int pktId=msg->getPktId();
        int flitIdx=msg->getFlitIdx();
        int schedulingPriority=msg->getSchedulingPriority();
        bool firstNet=msg->getFirstNet();
        int flits=msg->getFlits();
        int totalCount=msg->getTotalCount();

//EV<<"Scheduler sending flit to (Out) link"<<endl;

        send(msg, "out$o", 0);

//EV<<"Setting up 3 ns timer, the detect fault, if any"<<endl;

        popRouterFaulty= new NoCRouterFaultyMsg("Router
        Faulty");
        popRouterFaulty->setKind(NOC_RFAULTY_MSG);
        popRouterFaulty->setType(type);
        popRouterFaulty->setVC(vc);
        popRouterFaulty->setSrcId(srcId);
        popRouterFaulty->setDstId(dstId);
        popRouterFaulty->setPktId(pktId);
        popRouterFaulty->setFlitIdx(flitIdx);
        popRouterFaulty->setSchedulingPriority(schedulingPrio
        rity);
        popRouterFaulty->setFirstNet(firstNet);
        popRouterFaulty->setFlits(flits);
        popRouterFaulty->setTotalCount(totalCount);
        popRouterFaulty->setSynpseId(synpseID);
        for (int i=0; i<totalCount;i++)
          {
            int outPort=msg->getSynpOutPort(i);
              popRouterFaulty->setSynpOutPort(i,outPort);
          }
        for (int i=0; i<totalCount;i++)
          {
            int routerIndex=msg->getRouterIndex(i);
              popRouterFaulty->setRouterIndex(i,routerIndex);
          }
            simtime_t newTime1=simTime()+0.000000003;
            scheduleAt(newTime1, popRouterFaulty);

//EV<<"This module is used to block particular ports"<<endl;
```

Appendix B   187

```cpp
//EV<<"Informs inPort to initiate a new workaround
synapse"<<endl;

void SchedAsync::handleRouterFaultyMsg(NoCRouterFaultyMsg* msg)
{
        cModule* curRouter = getParentModule()->
        getParentModule();
        cModule* curPort = getParentModule();
        int srcId=msg->getSrcId();
        int dstId=msg->getDstId();
        int totalCount=msg->getTotalCount();
        int synpseID=msg->getSynpseId();
        faultyPktId=msg->getPktId();

//EV<<"If the router port is detected faulty as second
time then only send flit not initiate new synapse as other
workaround synapse is already constructed or is under
construction"<<endl;

        if (oldRouter!=curRouter->getIndex())
        {
            sourceInform=new NoCInformSourceMsg("Inform
            Source");
            sourceInform->setKind(NOC_INFORMSOURCE_MSG);
            sourceInform->setSrcId(srcId);
            sourceInform->setDstId(dstId);
            sourceInform->setTotalCount(totalCount);
            sourceInform->setSynpseId(synpseID);
        for (int i=0; i<totalCount;i++)
          {
            int outPort=msg->getSynpOutPort(i);
            sourceInform->setSynpOutPort(i,outPort);
            }
            for (int i=0; i<totalCount;i++)
            {
              int routerIndex=msg->getRouterIndex(i);
                sourceInform->setRouterIndex(i,routerIndex);
            }
            send(sourceInform,"sched_link$o");

//EV<<"Deleting flit from qSched as this flit has already sent
to the inPort"<<endl;

            NoCFlitMsg* flitFaulty = (NoCFlitMsg*)
            qSched.pop();
            delete(flitFaulty);
        }
```

```
            else //on second faulty router detection, just
            send the flit to inPort
            {
                NoCFlitMsg* flitFaulty = (NoCFlitMsg*)
                qSched.pop();
                flitFaulty->setLostFlit(true);
                send(flitFaulty, "sched_link$o");
            flitFaulty=NULL;
                delete (flitFaulty);
        }
```

//EV<<"If the router port is detected faulty as second
time then only send flit do not initiate new synapse as
other workaround synapse is already constructed or is under
construction"<<endl;

```
                oldRouter=curRouter->getIndex();
                cancelAndDelete(msg);
}
```

Sprouting Algorithm

```
if (portNumber == 0) {
        int northPort = curRouter->par("northPort_W");
        if (northPort == 0)
    {
```

//EV<<"If the router port is detected as first time
faulty"<<endl;

//EV<<"Then inform inPort to initiated new workaround
synapse"<<endl;

```
        if (tagRouterFaulty==false)
        {
            tagRouterFaulty=true;
            sourceInform = new NoCInformSourceMsg("Inform
            Source");
            sourceInform->setKind(NOC_INFORMSOURCE_MSG);
            sourceInform->setSrcId(srcId);
            sourceInform->setDstId(dstId);
            sourceInform->setTotalCount(totalCount);
            sourceInform->setSynpseId(synpseID);
              for (int i = 0; i < totalCount; i++)
              {
                int outPort = msg->getSynpOutPort(i);
                sourceInform->setSynpOutPort(i, outPort);
              }
```

Appendix B

```cpp
        for (int i = 0; i < totalCount; i++)
        {
          int routerIndex = msg->getRouterIndex(i);
          sourceInform->setRouterIndex(i, routerIndex);
        }
          send(sourceInform, "sched_link$o");
    }
        faultySynpseID = synpseID;
        msg->setLostFlit(true);
```

//EV<<"Sending flit to inPort to save in Queue to send over workaround synapse"<<endl;

```cpp
        send(msg, "sched_link$o");
```

//EV<<"Incrementing credit counter by one so that the current port can receive flits from other ports of the current router"<<endl;

```cpp
        creditPop = new NoCPopCreditMsg();
        creditPop->setKind(NOC_POPCredit_Msg);
        creditPop->setVc(msg->getVC());
        scheduleAt(simTime() + 0.000000002, creditPop);
      return;
    } else {
```

//EV<<"If the router port is detected faulty a second time then only send flit do not initiate new synapse as other workaround synapse is already constructed or is under construction"<<endl;

```cpp
        if (faultySynpseID == synpseID)
        {
            msg->setLostFlit(true);
            send(msg, "sched_link$o");
            creditPop = new NoCPopCreditMsg();
            creditPop->setKind(NOC_POPCredit_Msg);
            creditPop->setVc(msg->getVC());
            scheduleAt(simTime() + 0.000000002, creditPop);
        return;
        }
    }
```

Index

A

accepted traffic (flit/cycle/node), 103
Advanced Microcontroller Bus Architecture (AMBA), 30
algorithm 1, 72
algorithm 2, 73
algorithm 3, 74
ant colony optimization (ACO), 49
area and power consumption, 33
artificial bee colony (ABC) algorithm, 50
artificial immune system, 49
asynchronous, 21, 22, 23, 32, 90, 109, 116, 131, 134, 136, 146
autonomic network on chip using the biological immune system, 52
autonomous error tolerant (AET) architecture, 52

B

bandwidth, 1, 4, 5, 6, 9, 10, 13, 14, 20, 22, 24, 30, 34, 35, 36, 37, 59, 64, 79, 81, 84, 88, 89, 90, 91, 92, 94, 96, 97, 98, 99, 106, 107, 109, 110, 118, 119, 125, 132, 149, 152, 158, 162, 163, 175
bat algorithm (BA), 51
BE, 7, 36, 37, 38, 56, 74, 75, 76, 77, 84, 85, 86, 88, 91, 94, 97, 98, 99, 100, 101, 102, 103, 104, 105, 106, 107, 110, 114, 116, 117, 118, 120, 121, 124, 126, 127, 134, 138, 144, 145, 146, 150, 155, 158
best effort, 6, 36, 56, 91, 110
bio-inspired, 4, 6, 7, 38, 49, 50, 51, 52, 53, 55, 56, 57, 59, 60, 62, 64, 72, 73, 74, 82, 84, 86, 87, 89, 90, 91, 94, 95, 96, 97, 98, 99, 100, 101, 103, 105, 106, 109, 110, 111, 185
bio-inspired online fault detection in the NoC interconnect, 52
bio-inspired self-aware NoC fault-tolerant routing algorithm, 52
biological brain characteristics, 55
brain, 4, 38, 51, 52, 55, 56, 57, 109
broadcast communication, 28, 35
buffering mechanisms, 28
buffer management techniques, 19
bus, 1, 29, 30, 31, 32, 33, 34, 35, 109, 114, 115, 123, 137, 141, 149, 150, 153, 154, 155, 156, 157, 161, 162, 163, 165, 166, 167

C

cat swarm optimization, 50
channel, 9, 10, 11, 16, 17, 19, 20, 22, 29, 34, 35, 36, 37, 76, 79, 81, 82, 83, 84, 86, 89, 97, 104, 110, 117, 120, 123, 126, 128, 136, 140, 143, 145, 148, 151, 152, 153, 156, 158, 160, 161, 162, 163, 164, 165, 166, 169, 173
circuit switching, 9, 10, 13, 17, 110, 114, 115, 116, 117, 119, 124, 128, 129, 130, 131, 132, 134, 136, 139
clocking mechanism, 21, 32, 34, 35, 114
cobweb, 51
combined BE-GT, 7, 37, 38, 74, 84, 85, 88, 91, 98, 99, 100, 101, 102, 103, 104, 105, 106, 107, 110
congestion, 1, 2, 5, 11, 13, 14, 15, 19, 24, 25, 26, 29, 34, 35, 76, 88, 109, 118, 119
connectionless mechanism, 7
connection-oriented, 5, 6, 9, 10, 13, 100
connection-oriented stochastic routing, 5, 13
connection type, 26, 35, 114
contention free routing, 20, 81
core, 1, 6, 31, 34, 62, 67, 74, 83, 88, 114, 117, 118, 120, 132, 133, 135, 139, 141, 146, 159, 163
COSR, 5, 13
credit-based flow control, 6, 7, 75, 110, 123, 141
credit based technique, 19
crossbar, 1, 24, 25, 62, 109, 114, 117, 118, 119, 122, 123, 124, 125, 127, 129, 131, 132, 133, 138, 139, 144, 169
cuckoo search (CS), 50
cuttlefish algorithm (CFA), 51

D

data communications, 1, 109
deadlock, 4, 5, 11, 12, 13, 14, 15, 16, 17, 18, 20, 34, 76, 83, 88, 127, 145
destination, 1, 2, 3, 6, 9, 10, 11, 12, 13, 14, 15, 16, 18, 19, 20, 23, 24, 26, 27, 28, 29, 34, 35, 36, 52, 57, 59, 64, 65, 67, 68, 69, 70, 71, 73, 74, 75, 76, 77, 80, 84, 85, 86, 87, 90, 92, 93, 94, 98, 99, 102, 103, 105, 109, 110, 154
deterministic, 3, 4, 5, 6, 11, 12, 17, 38, 99, 101, 109, 114, 115, 116, 118, 119, 120, 121, 123, 125, 126, 127, 128, 129, 130, 134
devices, 1, 2, 21, 22, 30, 33, 90, 109
dimension order routing, 12, 122

Index

directed flooding, 13
DOR, 12, 15

E

end-to-end latency, 4, 98, 124, 135, 165
epidemic spreading, 49

F

fat tree, 24, 35, 114, 117, 120, 128, 134, 137, 159
faults, 1, 2, 4, 15, 17, 38, 51, 52, 57, 58, 59, 72, 74, 75, 79, 89, 90, 92, 94, 95, 96, 97, 98, 99, 100, 101, 102, 103, 104, 105, 106, 109, 110, 148, 159, 171, 173
fault tolerance, 1, 2, 4, 5, 12, 13, 15, 17, 24, 35, 38, 71, 109, 148, 152, 165
fault-tolerant, 1, 2, 4, 5, 6, 7, 11, 12, 13, 14, 15, 17, 38, 52, 55, 56, 57, 89, 100, 101, 103, 104, 105, 106, 109, 110, 185
fault-tolerant NoC using biological brain techniques, 52
firefly algorithm, 49
flits, 5, 76, 77, 80, 85, 155, 160, 186
flit size, 34, 35, 91, 148, 151, 153, 161, 163, 165, 166, 169
flow control technique, 91
flower pollination algorithm (FPA), 50
force-directed wormhole routing, 5, 14
framework, 7, 59, 89, 90, 91, 123
frequency and technology, 32
fully adaptive, 3, 4, 5, 6, 11, 14, 15, 38, 109, 143, 171
future work, 109, 110

G

globally asynchronous and locally synchronous, 21
GT, 7, 36, 37, 38, 56, 74, 75, 79, 84, 85, 86, 88, 91, 97, 98, 99, 100, 101, 102, 103, 104, 105, 106, 107, 110, 116, 124, 136, 145, 146, 158
guaranteed throughput, 6, 7, 36, 56, 91, 110

H

Harris hawks optimization (HHO), 51
head of line, 29, 79, 110
heterochronous, 21, 23, 32
hierarchical-routing-table-based FTDR algorithm, 15

I

implementation platforms, 28
injection rate, 9, 89, 90, 91, 97, 99, 100, 106, 127, 128, 131, 138, 139, 141, 142, 145, 170, 172

inPort, 59, 60, 61, 62, 64, 71, 72, 73, 74, 75, 76, 79, 83, 84, 85, 86, 155, 185, 187, 188, 189
input queue, 29, 148, 149, 150, 151, 158, 159, 166, 168, 169, 170, 172
interconnect, 1, 3, 6, 11, 20, 21, 22, 25, 29, 30, 34, 35, 38, 52, 57, 58, 59, 60, 64, 67, 68, 70, 71, 72, 73, 74, 75, 78, 81, 88, 89, 91, 94, 95, 102, 103, 109, 110, 114, 117, 123, 127, 128, 130, 144, 148, 151, 158, 160, 164, 171, 172, 186
Internet Protocol (IP), 1

K

killer whale algorithm (KWA), 51

L

latency, 1, 4, 5, 11, 12, 15, 17, 18, 20, 22, 24, 25, 26, 28, 29, 34, 35, 36, 51, 53, 59, 90, 91, 92, 93, 94, 95, 96, 97, 98, 99, 100, 107, 110, 114, 115, 116, 117, 118, 119, 120, 122, 123, 124, 126–174
link, 9, 10, 11, 15, 19, 20, 21, 24, 25, 29, 59, 60, 72, 73, 74, 76, 79, 81, 84, 110, 114, 186
livelock, 5, 11, 13, 16, 17

M

mesh, 1, 24, 25, 28, 34, 51, 74, 90, 91, 114
mesochronous, 21, 23, 32, 116
message, 1, 10, 15, 17, 26, 27, 32, 35, 51, 77, 86, 114, 185
minimal path routing, 12
multicast connections, 27

N

nanotechnology, 2
narrowcast, 26, 27
Network on Chip (NoC), 1, 2, 9, 26, 35, 51, 56, 89, 90, 109, 124
NI, 1, 11, 20, 21, 29, 31, 66, 75
N-random walk, 5, 13

O

octagon, 24, 25, 115, 129, 137, 149, 166
on-chip communication, 1, 109
Open Core Protocol (OCP), 31
open-source, 25, 26
OutPortCalc, 59, 65, 82
output queue, 29, 148, 150, 151, 152

P

packet, 1, 4, 6, 9–21, 28, 34–37, 52, 58, 59, 60, 61, 62, 64–86, 88–92, 94, 97, 99, 100, 106, 109–111, 114

Index

193

partial adaptive, 3–6, 11, 15, 16, 17, 38, 109
performance evaluation parameters, 91
PEs, 1, 6, 9, 10, 11, 14, 19, 20, 21, 22, 24, 25, 26, 27, 29, 30, 31, 36, 38, 74, 75, 106, 109, 110, 146
planar adaptive routing algorithm, 16
plesichronous, 21, 23, 32, 118
power, 5, 13, 14, 15, 16, 17, 19, 20, 21, 22, 24, 28, 30, 33, 34, 35, 53, 88, 114
processing elements, 1, 9, 52, 109

Q

quality of services (QoS), 110

R

reliable, 1, 3, 6, 11, 19, 52, 88, 133, 164
research framework, design and parameters, 89
resources, 1, 2, 3, 9, 11, 12, 13, 17, 20, 36, 75, 97, 102, 109, 110, 119, 140
ring, 1, 4, 25, 115, 119, 131, 134, 135, 140, 145, 153, 169
round-robin arbitration, 10, 114, 117, 121, 122, 123, 137, 146
router ports and bus width, 34
routers, 1, 2, 5, 6, 9, 10, 14–22, 24, 25, 29, 36, 52, 59, 61, 62, 64, 65, 67, 68, 73, 74, 75, 76, 77, 79, 80, 81, 85, 88, 91, 94, 106, 109, 110, 111, 123, 124, 125, 128, 144, 167
routing algorithm, 1–7, 10–17, 24, 24, 34, 35, 38, 51, 52, 67, 68, 77, 91, 94, 96, 98, 104, 109, 111, 164
routing table, 4, 5, 6, 14, 15, 16, 88, 128
R-Software, 90

S

saturation point, 35, 93, 94, 103, 105, 106, 110, 149, 155, 159, 166, 167, 168, 169, 170, 173, 174
scalable bio-inspired fault detection unit, 51
scheduler, 29, 59, 60, 61, 62, 64, 71, 72, 73, 74, 75, 76, 79, 81, 82, 83, 84, 85, 86, 186
self-adapt, 6, 52, 56
self-heal, 6, 38, 52, 56
self-optimize, 6
services, 6, 9, 14, 36, 37, 38, 56, 74, 75, 76, 79, 84, 85, 86, 88, 91, 97, 103, 104, 106, 107, 110, 118, 124, 156
shared virtual channel control technique, 19, 20
simple connection, 26
source routing, 5, 11, 14, 124

spidergon, 24, 25, 121, 129, 132, 139, 153, 160
SpiNNaker communication, 52
sprouting, 6, 38, 52, 56, 57, 58, 59, 72, 73, 91, 97, 98, 99, 100, 101, 102, 103, 104, 105, 109, 188
star, 23, 24, 115, 118, 122, 124, 126, 129, 139, 144, 146, 152, 174
stochastic, 3, 4, 5, 6, 11, 13, 15, 17, 38, 109
store and forward, 17, 18, 119, 131
subnets, 24, 25, 128, 141
summary, 7, 38, 88, 106
swarm intelligence algorithms (SIA), 49
synaptogenesis, 6, 38, 52, 56, 57, 59, 72, 91, 94, 98, 101, 102, 103, 104, 105, 109, 185
synchronous, 21, 22, 31, 32, 34, 53, 90, 109, 114

T

throughput, 4, 6, 7, 9, 10, 11, 12, 14, 17, 20, 24, 34, 35, 36, 51, 53, 56, 59, 81, 82, 90, 91, 93, 95, 96, 97, 101, 102, 103, 106, 107, 110, 114
time division multiplexing (TDM), 16, 79, 81, 83, 87, 91, 97, 98, 99, 100, 101, 102, 106, 110, 114
topology, 1, 24, 25, 34, 35, 51, 74, 90, 91, 114
torus, 1, 3, 24, 28, 51, 114

V

virtual channels, 9, 10, 11, 16, 17, 37, 76, 79, 81, 83, 86, 110, 117
virtual cut through, 17, 18, 125, 129, 131, 133, 134, 140, 142
virtual output queue, 29, 165, 172
Virtual Socket Interface Alliance (VSIA), 32

W

west first, negative first, north last, south last routing algorithms, 16
wormhole switching, 10, 14, 17, 18, 34, 89, 90, 91, 114

X

XY, 5, 12, 91, 96, 98, 99, 101, 102, 105, 117, 118

Y

year of proposal, 34

Z

ZYX, 5, 12